The Gnu's World

The publisher gratefully acknowledges the generous support of the General Endowment Fund of the University of California Press Foundation.

The Gnu's World

Serengeti Wildebeest Ecology and Life History

Richard D. Estes

UNIVERSITY OF CALIFORNIA PRESS

Berkeley · Los Angeles · London

University of California Press, one of the most
distinguished university presses in the United States,
enriches lives around the world by advancing scholarship
in the humanities, social sciences, and natural sciences. Its
activities are supported by the UC Press Foundation and
by philanthropic contributions from individuals and
institutions. For more information, visit www.ucpress.edu.

University of California Press
Berkeley and Los Angeles, California

University of California Press, Ltd.
London, England

Library of Congress Cataloging-in-Publication Data

Estes, Richard.
   The gnu's world : Serengeti wildebeest ecology and life
history / Richard D. Estes.
      pages cm
   Includes bibliographical references and index.
   ISBN 978–0-520–27318–4 (cloth : alk. paper) —
   ISBN 978–0-520–27319–1 (pbk. : alk. paper)
   1. Gnus—Tanzania—Serengeti Plain. I. Title.

QL737.U53E88 2014
599.64′59—dc23
                                          2013031723

Manufactured in the United States of America

23 22 21 20 19 18 17 16 15 14
10 9 8 7 6 5 4 3 2 1

To Runi

Without whose collaboration, companionship,
and support

conducting research in Africa over the past half
century would have been

far less rewarding or even possible

Contents

Acknowledgments

How can I possibly acknowledge all the people who deserve thanks for their encouragement, support, friendship, and hospitality over the 50+ years since I began research in Africa? Going over progress reports and papers written as long ago as the sixties, I realized that I had forgotten some friends and colleagues who helped in one way or another. My review served to fill most of the gaps, but, considering that my memory continues to diminish with age, inevitably I'll fail to include everyone. For which I can only ask indulgence.

As most of the years I've studied the wildebeest have been spent in the Serengeti ecosystem and involved the western white-bearded sub-species, the lion's share of governmental and nongovernmental organizations, research grants, collaborators, supporters, and friends who need to be acknowledged will be from that region of Tanzania. Times and places are given in the introduction. All told, I spent less than a year in the Masai Mara part of the ecosystem, between 1963 and 2007.

Research I did on all the other wildebeest populations, in Kenya, Tanzania, and the other range states of eastern and southern Africa, was carried out during visits to parks and reserves lasting from a few days to a week or two. I shall acknowledge the organizations and individuals who enabled the research to take place.

Research Grants, Institutional, Corporate, and Private Support

The National Geographic Committee for Research and Exploration has been most supportive, with grants for research on the wildebeest and other antelopes in the Serengeti ecosystem in 1963–65, 1973, 1975, 1977, 1978, 1982, and 1996–98. I am grateful to the following for their support: Tanzania Ministry of Agriculture, Forestry, and Wildlife (1963–64); New York Explorers Club (1962); National Science Foundation and Academy of Natural Sciences (research on Serengeti ungulates, 1979–81). International Union for Conservation of Nature (sponsorship, vehicle loaned by Nairobi HQ, 1996). World Wildlife Fund–US, British Airways Assisting Conservation, Environment Branch (three complimentary tickets, Boston-Nairobi, 1988, 1989); Safari Club International (Massachusetts Chapter, 1999). World Wildlife Fund–US. Smithsonian Institution Conservation and Research Center (reproductive physiology of the wildebeest, fieldwork, 2001–3); Paul Tudor Jones and the Grumeti Reserves (financial and logistical support of the captive wildebeest maintained there during the reproductive study, 2002–4).

A University of Cape Town Fellowship enabled me to carry out research on wildebeest in southern African parks and reserves in 1986.

Paul Khurana, Rupert Ingram, Gerry Mann, Lisa Gemmill, David Van Vleck, and Roger Wood made generous contributions to my research.

Tanzania

For permission to conduct research in Tanzania, I thank the Tanzania Wildlife Research Institute and the Commission for Science and Technology. The Tanzania Game Department (Bruce Kinloch), chief conservators, from Henry Fosbrooke and Ole Saibull (1963–65) to Emmanuel Chausi (1995–2002), permitted and facilitated the research I carried out in the Ngorongoro Conservation Area. The directors general of Tanzania National Parks enabled me to collect data on wildebeest and other wildlife in Serengeti, Tarangire, and Manyara National Parks, wherein park wardens and rangers provided much-needed assistance. The College of African Wildlife Management (Mweka) carried out semiannual game counts in Ngorongoro. I thank the Wildlife Division for permission to stay in the Grumeti Reserves.

Collaboration and Cooperation with Researchers and
Wildlife Managers

The following individuals and organizations helped my research by providing assistance, guidance, and/or information.

Ngorongoro Conservation Area (NCA). A capture-marking operation
of yearling Crater wildebeest by warden David Orr and wildebeest
researcher Murray Watson enlarged my sample of known individuals
(1963–65 and 1973). John Goddard shared data on predation by a resident population of wild dogs (1964–65; Estes and Goddard 1967);
Fritz Walther shared observations of Crater gazelles (1964); Winston
and Lynne Trollope prepared a management plan for Ngorongoro
grasslands (1995 and 1998), including scheduled burning, assisted and
later implemented by Amiyo Tlaa, who also assisted my research. Daniel Deocampo shared his preliminary survey of the Crater's surface
waters (1999). Hans and Ute Klingel collaborated with marking operations. Bernard Kissui, Dennis Ikanda, and Ingela Janssen, monitoring
Crater lions as part of Craig Packer's ongoing Serengeti lion research,
responded to requests for information about conditions between my
trimestral visits (1997–2002 and later). George and Lori Frame provided information on wildebeest and conditions in the NCA while doing
research in Mpakai Crater in the early 1970s.

Serengeti National Park. Park wardens, from Myles Turner in the
1960s to Justin Hando in 2002, facilitated my research in various
ways, including help capturing and marking animals, permission to
follow the migration and record behavior of radio-collared gnus, and
access to park facilities that included stays at the Seronera rest house.
During visits to Serengeti National Park from 1979 on, I've enjoyed the
support extended to researchers by Markus Borner and the Frankfurt
Zoological Society (FZS), including Internet access, two-way radio
contact, vehicle repairs, flights on FZS aircraft, and briefings and meet-
ings. Other helpful colleagues/friends are Jack Ingram (darting, marking topi); Timothy Tear (wildebeest in Tourist Center ecological
exhibit); and George and Lori Frame, who hosted my 1977 stay in
Serengeti National Park. Martyn Murray and Patrick Duncan shared
information on differences in dietary and feeding specializations of the
Alcelaphini.

Voluntary Participants in Fieldwork

Ngorongoro. J. Pettit (1964), Kathryn Fuller (1973), Anna Estes (1997)

Serengeti. Dan Otte, Marina Botje, William Baker, David Van Vleck, Jessie Williams, Karen Nielsen, Lisa Gemmill, Anna Estes, Roger Wood.

Grumeti Reserves. Support of the wildebeest reproductive study (2002–4): fence construction, delivery of water and feed, access to clinic, fuel and auto repairs, day laborers, and game guards (Rian and Lorna Labuschagne). National Park veterinarians and Richard Hoare from the Research Center helped keep our animals healthy. Nicky Jenner and Penny Spierling assisted Allison Moss's dissertation research.

Eastern Masailand. I thank the wardens of Manyara, Tarangire, and Arusha National Parks for admission and assistance. For consultation and information on wildebeest in the parks and on the Manyara Ranch, at Lake Eyasi and the Wembere Plain, Lake Natron, the Simanjiro Plains, and Mkomazi, I thank Hugh Lamprey, Gil Child, Esmond Vesey-FitzGerald, Max Morgan-Davies, David Peterson, Herbert Prins, A.C. Brooks, Alan Rodgers, Institut Oikos (Valeria Galanti, Guido Tausi), David Moyer, Marc Baker, Charles Foley, Tom Morrison, and Malcolm Coe.

Mikumi National Park and Selous Game Reserve. Warden Brian Nicholson, GTZ (German Technical Cooperation Agency) (Rolf Baldus, Ludwig Siege).

Friends with Whom We Regularly Stayed

Arusha and environs: Andreas von Nagy, proprietor of the Mount Meru Lodge and Game Sanctuary, where I (later with family) regularly stayed when in Arusha, from 1963 to 1981; Elena Brooke-Edwards; Diana Cardoso and Bruce Russell (1996–2002); Annette and John Simonson (Tarangire Lodge); Tanganyika Game Trackers; Gibbs Farm (Margaret Gibb and managers).

The Peterson family's Dorobo Safaris has been my unofficial HQ in Arusha since the 1980s, where I've received all kinds of help, in addition to repairs and storage of my old Land Rovers between field sessions. Most of the safaris I organized were outfitted and led by Dorobo.

In and Near the Ngorongoro Conservation Area. Henry Fosbrooke; Frankfurt Zoological Society NCA establishment (Rian and Lorna Labuschagne, Johan and Paula Robinson, Pete Morkel); Carol and

Jonas Sorenson, David Bygott and Jeannette Hanby, Crater Lodge; Ndutu Lodge (Aadje Geertsema, Louise and Paul, and other lodge managers); Hugo van Lawick's Ndutu encampment.

Serengeti National Park. Places I often stayed on short visits (1997–2004): Craig Packer's lion house and Simon Mduma's house at the Serengeti Research Center).

Kenya

For permission to carry out research in parks and reserves, I thank the Kenya Wildlife Service and, before the merger of the Game Department and National Parks, Ian Grimwood and Merwin Cowie respectively.

Masai Mara National Reserve: my understanding of the wildlife and impacts of development benefited from the research of Morris Gosling, Jacob Bro-Jorgensen, Laurance Frank, Holly Dublin, Kay Holecamp and her students; Stephanie Dloniak; the Mara Conservancy; Richard Lamprey; and Joseph Ogutu.

Nairobi and Amboseli National Parks, Athi-Kapiti Plains: Kenya Wildlife Management Project (Robert Casebeer); Don Stewart, Bristol Foster, Morris Gosling, Jesse Hillman, Ian Cowie, Stuart and Jeanne Altmann, Cynthia Moss, David Western, Alan Root, Harvey Croze, David Hopcraft, Mike Rainey, Peter Greenway, Mike Norton-Griffiths, Mark Stanley Price, Steve Turner, African Wildlife Foundation, IUCN Langata HQ, Tom Butynski, Richard Kock.

For the hospitality offered by friends during the times I (with and without family) spent in Nairobi: Elena Brooke-Edwards, Yolanda Brooke-Edwards, Peter Beard, Esmond and Chrysee Bradley Martin, Iain and Oria Douglas-Hamilton, David Hopcraft, Barbie Allen.

Eastern and Southern Africa

Governmental and nongovernmental organizations and colleagues/friends who assisted our wildebeest research in the other range states.

Zambia

William Bainbridge, Chief Game Officer, National Parks; Frank Ansell, Leslie Robinette, Richard Jeffery, P. Berry, Ed Sayer, Petri Viljoen.

Zimbabwe

1965: Reay Smithers, Thane Riney, A. Mossman, Ray Dasmann, H. H. Roth, Vivian Wilson, Alan Savory, Wankie National Park (E. Davison, Derek Williams); Jeremy Anderson, Sarah Clegg (Malilangwe), Derek and Sarah Solomon.

Botswana

Reay Smithers, Bruce Kinloch, Rowan Martin, L. Tennant, D. Williamson, B. Mbano, Clive Spinage, Karen Ross.

Namibia

Department of Nature Conservation; Ken Tinley, Hymie Ebedes, Garth Owen-Smith, Hu Berry, William Gasaway, Philip Stander.

Angola

J. Crawford Cabral, Brian Huntley, Kissama Foundation, IUCN Angola.

South Africa

National Parks Board; KwaZulu-Natal Parks (Rudi Bigalke, Peter Goodman, U. de V. Pienaar; Brian Huntley, Norman Owen-Smith, Petri Viljoen, Ian White, Salomon Joubert, Gus Mills, Daryl Mason, Jeremy Anderson; Wouter Van Hoven (Pretoria University); Hluhluwe Game Reserve (R. I. G. Attwell); Willem Pretorius Reserve (W. von Richter, Savvas Vrahimis); Todd Kaplin (Wildlife Campus), Russel Friedman, Eleanor-Mary Cadell.

Mozambique

José Tello, Ken Tinley, Reay Smithers, J. Augusto Silva, W. von Alvensleben, Gregory Carr (Gorongosa National Park).

Contributors to This Book

I am very much indebted to Mark Davis for the cover photo; Ken Coe for contributing his outstanding photos of the wildebeest's tribe; Roger

Wood for increasing the resolution of many photos to publication standards; Grant Hopcraft for preparing and providing maps; Neil Stronach for colorful and historical information included in chapter 12; Martyn Murray and Hu Berry for copies of their PhD dissertations; Kevin Kirkman for a copy of Amiyo Tlaa's thesis; Paul Reillo, Sy Montgomery, and Runi Estes for reading the entire manuscript and cheering me on; and Dennis Herlocker, Martyn Murray, Lyndon Estes, Anna Estes, Allison Moss Clay, Mathew Brown, David Banks, Robin Reid, Fiona Marshall, Patrick Duncan, John Fryxell, and Serengeti Watch for reviewing specific chapters.

For continuing assistance in keeping my computer operating, Bill Kennedy, coauthor of a definitive guide to HTML (Hypertext Markup Language) and a neighbor, has been most helpful. Trivelore Ragunathan, Michigan State University professor of computer science, designed and ran computer analyses of my distribution data as a Harvard graduate student (1998 and later; Estes, Raghunathan, and Van Vleck 2008). Jon Atwood of Antioch New England Graduate School applied his GIS and mapping expertise to plotting the distributions of the five species of ungulates that I compared in Ngorongoro and the Serengeti (1997–2002). Anna Estes not only participated in my research but also continued sampling Crater and Serengeti ungulates for several months on her own (1997) (Estes, Atwood, and Estes 2006). In the Grumeti Reserves, the IT department rescued the hard disk when my computer crashed and provided other much-needed technical support;

The Ernst Mayr library in the Harvard Museum of Comparative Zoology has been my main source of publications dating back to 1966. Mary Sears, Emily Dark, and Runi used these resources to prepare the bibliography.

At the University of California Press I want to thank Blake Edgar, senior sponsoring editor; Merrik Bush-Pirkle; Sheila Berg; Rich Nybakken; Lynn Meinhardt; and Francisco Reinking.

Introduction

The Author's Fifty-Year History of
Wildebeest Research

My purpose in writing about the natural history of the wildebeest is twofold: to give the antelope that once dominated the plains of eastern and southern Africa a book all to itself and to repay my debt to the animal that I have studied off and on for half a century while privileged to live in World Heritage Sites and International Biosphere Reserves.

How did I single out the wildebeest from among all the other teeming plains game? I didn't plan it that way.

I had dreamed of living among the large mammals on the African savannas from the age of ten, when I was exposed to Carl Akeley's life-like dioramas at the American Museum of Natural History in New York. I tried from the time I graduated from college in 1950 to find employment in an East African park. But another decade passed before the dream came true, and by then I was already in my thirties.

Meanwhile, I had spent four years working as a journalist-photographer on *Yankee Magazine* in Dublin, New Hampshire, three years writing a social and natural history of the Atlantic East Coast that the publisher canceled within sight of the finish line, and two years on a wildlife survey of Burma.

Burma was somewhat of a detour from my primary objective, but en route I had the opportunity to spend a summer at the Max Planck Institute for Behavioral Physiology in Bavaria, where I studied with Konrad Lorenz and met other famous biologists, including Ernst Mayr, Carl von Frisch, and Oxford's Niko Tinbergen, who later shared with Lorenz the Nobel Prize for founding the field of ethology.

I had been drawn to ethology since reading Lorenz's popular book, *King Solomon's Ring,* four years out of college. That summer in 1958 convinced me that my path was to go back to college and work for a PhD in ethology. My dissertation research, naturally, would entail the study of African large mammals. But all that had to wait until after the Burma wildlife survey.

When I finally arrived in Tanzania in October 1962, I was a Cornell graduate student, and I had secured permission to do my dissertation research in Ngorongoro Crater, often described in travel literature as "the eighth wonder of the world." My first view of this immense bowl in the Crater Highlands made the claim almost believable. Chugging up the steep, winding road from the entrance gate to the Ngorongoro Conservation Area (NCA) through lush montane forest in the 1958 Land Rover pickup I bought for fifty pounds ($150) in Nairobi, I was unprepared for the view that opened at my feet when I finally emerged on the Crater rim. In a landscape dominated by mountains 10,000 to 15,000 feet high, I gazed into an amphitheater 10 by 12 miles in diameter, 1,000 feet below where I stood. The floor was an open plain and tan, in contrast to the surrounding montane grassland and forest, for it was late in the six-month dry season. Here was a perfect microcosm of the Serengeti plains that began at the western foot of the Crater Highlands (fig. 0.1). Ngorongoro could be likened to a vast diorama of the East African plains populated with living animals.

But where was all the teeming wildlife? From my viewpoint, all I could see through the afternoon haze were little dots and dark clusters that might be animals, or only bushes. I had to look through my 8×32 binoculars to resolve the nearest clusters into wildebeest, zebras, and gazelles, with a few elephants visible in a swamp close to the inner wall. The single dots I had seen turned out to be wildebeest, whose dark color made them stand out from the bleached grass. Why, I wondered, would these individuals, members of a species whose habit of gathering in dense concentrations proved they were highly sociable, isolate themselves like this? Could they be defending territories?

Oddly enough, the first behavioral observation I made as I gazed into the void of Ngorongoro Crater became the subject of my doctoral dissertation: the territorial behavior of the wildebeest.[1] But the wildebeest was only one of the five hoofed mammals (ungulates) I intended to study. My grant proposal was to carry out the first multispecies com-

FIGURE O.I. The west side of Ngorongoro Crater, where wildlife exits and enters.

parative behavioral study of associated African ungulates. The five I had chosen were the wildebeest, the zebra, Grant's and Thomson's gazelles, and the eland.

The practical difficulty of studying five species at once in equal depth could explain why no one had tried it before. Anyway, I quickly learned that two species on my list had to be deleted. A German scientist from the Serengeti Research Institute was already studying the zebra population and had no wish for a collaborator. And it quickly became clear that the eland was too peripatetic to keep under observation. A nomadic species with no territorial ties, herds of this largest antelope moved in and out of the Crater and could not be followed while they ranged the surrounding montane grasslands. The use of radio-tracking collars was still in the future. So were immobilizing drugs whose effects could be reversed. Any hope of marking eland had to be forsworn after the first attempt to capture one by the warden, David Orr, and me, using succinylcholine chloride delivered in a dart fired from a CO_2-powered rifle, killed a magnificent bull. The memory is still painful. That left me with three antelopes on my plate, so to speak.

Nearly two months passed before I was outfitted to live independently on the Crater floor. Meanwhile, I stayed in a house on the south rim as the guest of Henry Fosbrooke, the British civil servant of the Tanganyika Colonial Administration who became the first Conservator after the NCA was created in 1959. I commuted to the Crater floor by day, via the breathtaking Lerai descent road that traversed the inner wall in a series of switchbacks.

My campsite was beneath a grove of huge fig trees bordering the Munge River, a muddy stream carved into the Crater floor by water flowing from an adjacent higher crater, Olmoti, into little Lake Makat (a.k.a. Lake Magadi = Soda Lake; see map 0.1). It was an ideal location for my research, way out in the open grassland, yet inconspicuous to passing tourists. Here I was within sight, sound, and smell of some 20,000 ungulates. For the first six months I lived in an old British army officer's tent, served by one old and one young Mbulu man from Karatu, the principal town of this tribe on the southeastern edge of the NCA, whom I hired as cook and waiter, respectively. A game scout the warden assigned to assist me completed our small encampment.

Leaving aside animal visitors—hyenas, lions, jackals, an occasional rhino—individuals and groups of Masai often stopped by en route to and from their settlements *(manyattas)* (fig. 0.2). Several hundred of these pastoralists with several thousand cattle still lived inside the Crater during the 1960s. My visitors were motivated not only by curiosity about the *mzungu* (white person or European) in their midst but also by the hope of begging lifts to their destinations. My willingness to give rides on my way to or from NCA headquarters on the rim raised their expectations. It became more than a little frustrating when hitchhikers would walk up to my vehicle while I was doing research and ask me to take them to their destination. Why go on foot when there was a possible ride near at hand?

One reason my campsite was inconspicuous to passing tourists was the situation of my tent and the cooking tent fly in an overflow channel created in the distant past, several feet lower than the surrounding plain. One morning in early February 1963, after heavy rains, I awoke to find that the Munge had flooded and filled the overflow channel (fig. 0.3), forcing us to abandon camp for a week.

Oddly enough, that was the only time I was flooded out during the three years I lived there, including two months in 1973. Yet the rains in 1963–65 were so heavy that Lake Makat more than doubled in size, flooded and killed the nearest part of Lerai Forest, and turned a kilometer-wide strip of the plain into a marsh that I had to cross every time I went to NCA headquarters. The roads near the lake and through Lerai

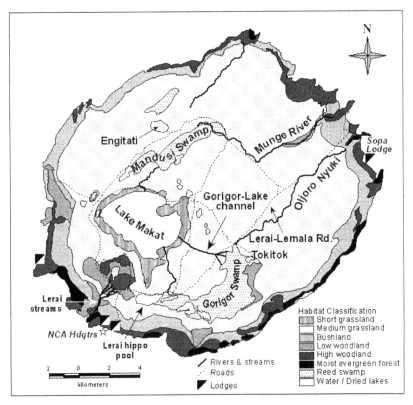

MAP 0.1. Ngorongoro Crater showing main landmarks and habitat diversity. Reprinted from Estes, Atwood, and Estes 2006, fig. 2.

Forest were underwater for months during and after the rainy seasons. Getting stuck was a course hazard. Once I spent the night in the Land Rover. Another time, accompanied by my camp crew, I walked the three miles home in the pitch dark; we kept up a conversation in hopes no rhino would take umbrage at our passing.

That I was allowed to live and work in Ngorongoro Crater for two and a half years seems in retrospect the most amazing good fortune anyone could have. Perhaps nowhere in Africa was there another place that combined so much scenic beauty with such an abundance of plains wildlife: some 25,000 large animals in a grassy bowl 10 to 20 miles wide from rim to rim. The floor of the Crater, at 6,000 feet, had a perfect temperate climate averaging about 65 degrees; there were no tsetse flies, no malarial mosquitos, and few other biting insects or poisonous

FIGURE 0.2. Masai settlement *(manyatta)* on the Crater floor, 1964.

FIGURE 0.3. Campsite on the Munge stream, after the big flood of 1963.

FIGURE 0.4. Same campsite after putting up the 12 x 18 ft. cabin where I lived from 1963 to 1965. Photo by J. Pettit.

snakes. It was altogether more pleasant on the floor than on the rim, 1,500 to 2,000 feet higher, where the offices and houses of the Ngorongoro Conservation Area Authority (NCAA) and the Crater Lodge were situated. Up there it was much colder and more often than not closed in by mist, which accounted, however, for a lovely cloud forest clothed in moss, lichens, and orchids.

By the middle of 1963 I had obtained permission to replace my tent with a one-room wooden cabin (fig. 0.4). It was prefabricated in clapboard sections that I brought down in the back of my pickup, along with 2-by-2-square-foot cement squares purchased from builders adding to the Crater Lodge. When my workers finished laying the floor, bolting the wall sections together, and putting on a tin roof (later muffled with thatch grass), I became the first European to live on the Crater floor since 1928, when Capt. Harry Hurst, ex-British army, was permitted to occupy the farm created by Adolf Siedentopf when Tanganyika was a German colony.[2]

Among all the memories of my time in the Crater, encounters with a leopard soon after moving into my new shack stand out most vividly. Here's an excerpt from my unpublished account of my early Crater years, titled "Life in a Game Paradise."

I was eating a solitary dinner in my cabin on the floor of Ngorongoro Crater when I heard, above the mumble of the stream and the hiss of the pressure lamp, a scratching noise just outside the window. Without moving from the chair, I reached to the desk for a flashlight, pushed the hinged window silently outward and shone the light up the trunk of one of the huge fig trees surrounding the house. There in the first fork, not 20 feet away, crouched a leopard. It looked directly into the light for several seconds, great yellow-green eyes unblinking. Then it turned and walked out along a branch to my observation platform, jumped onto it, and lay down, out of sight. It was still ensconced on the platform when I went to bed.

This was the first of many visits the leopard, a beautifully marked female in her prime, paid to my camp during the next six weeks. No leopard had come during the previous year and a half, and none came afterward. She must have taken up a temporary beat along this part of the stream, having wandered down from the wooded gorges of the rim via the belt of trees that lines the watercourse and provides the only cover in this part of the Crater grassland.

The platform made not only a comfortable bed, but a very satisfactory table on which to consume her meals in peace. They consisted mainly of jackals, which she caught with apparent ease as they came foraging near camp or prowled close to streamside thickets where she crouched in ambush. She brought eleven to the tree in the course of some fifteen visits, sometimes two in a night. After eating her fill she would curl up like a dog on a wildebeest skin I had put down for my own comfort and often sleep out the night there. It was amazing with what aplomb she treated me, Ami the houseboy, and Ama the cook. It didn't bother her at all if we stood on the porch and shone a light on her as she lay stretched on the platform or along a limb looking out at the plain. Nor did Ami's frequent trips between the kitchen and the cabin make her at all nervous. It took him a little longer, though, to get used to the idea of walking in the dark within 20 yards of a leopard. For that matter, none of us ever reached the point of walking beneath the tree while she was in it. This somewhat curtailed our movements and put my Land Rover, parked almost directly under the platform, out of reach. But it seemed a small price to pay for the privilege of being visited by such a magnificent and normally elusive animal. Besides, she came only at night and always left soon after dawn to lie up somewhere along the stream through the day, where I could never determine.

Sometimes, though, the noises that came from the platform were gruesome to hear, for the jackals that she brought were not always dead. While usually little could be seen from below, I once caught a glimpse of a jackal crawling dazedly about the platform. The leopard appeared to be playing with it exactly as a cat plays with a mouse. Next I heard a crunch followed by a blood-curdling shriek from the jackal. It did not pay to dwell in the imagination on this and similar scenes, and in a way I was thankful to be able to see so little.

Apart from jackals, the leopard caught and dispatched two adult male Grant's gazelles during her stay and dragged them into camp. They were members of a bachelor herd that frequented the area and had formed the unfortunate habit of browsing near the stream at night. Weighing about 160 pounds apiece, they were evidently too heavy for the leopard to take up the tree, so she left them beside my car, hardly touched. She ate part of the first one on the following night, and this time scratched twigs and leaves over the remains before leaving.

The trusting leopard almost lost the second Grant to Hans and Uta Klingel, the German zoologists from the Serengeti National Park who were studying zebra behavior in the Crater. When they called at my camp while I was on a trip to Arusha and Ami showed them the buck the leopard had brought in the night before, the sight of all that good meat going to waste was too much for them. So they butchered it on the spot, packed the quarters neatly in a box, and stored it away in the back of their Land Rover pickup. Returning the same afternoon, I met them on the road and invited them to join two other visiting scientists and me for dinner. While we were eating, the leopard arrived and, undeterred by all the noise, jumped into the back of the Klingels' Land Rover and lifted out the box. We found it on the ground as they were taking their leave, minus most of the meat. I like to think of the leopard indignantly reclaiming its rightful property; but of course it may have been simply thievery on her part, too. A week later an opportunity came to recompense the leopard for the Klingels' lèse-majesté, when a large Grant buck with a broken leg was put out of its misery. It took five of us to hoist the carcass up to the platform and secure it by wedging the horns between two branches. That night I heard a twig snap just as I was sitting down to dinner. As I looked out the window I saw the Grant rising smoothly above the level of the platform, where the leopard proceeded to dine while I ate my dinner at a separate table. Shortly after dawn the next morning I was awakened by a thumping on the platform and, looking out, saw the leopard walking along a branch carrying the gazelle, which was still largely intact. Evidently her grip was not secure, and she lowered it until its horns were caught on the limb, then let go to shift her hold. But it fell to the ground with a loud thud. In the next instant the startled leopard had bounded to earth and run up the path to my outhouse. There she turned and stood hesitating for a minute. But by now it was almost full daylight, the sound of voices was coming from the tent, and she thought better of it. Obedient to her presumed wishes, I proceeded to cover the carcass with leaves. To no purpose, however; she did not come back again for a long time.

No later African sojourn ever topped the experience of spending two and a half years in such a wildlife utopia. And even better, I happened to meet a beautiful young Austrian late in December 1963 who became my bride a year later (fig. 0.6).

To answer the question, finally, why I chose to focus on the wildebeest, they were my nearest neighbors and by far the dominant herbivore in the Crater. There were around 12,000 of them, plus or minus some thousands between the wet and dry seasons. Zebra, numbering

FIGURE 0.5. Releasing anesthetized and individually marked wildebeest in nearby study area. Photographer unknown.

about 4,000, came second in biomass (weight), followed by some 3,000 Thomson's gazelles and 1,500 Grant's gazelles. Buffalo, which later became the dominant herbivore, were represented by only dozens of bulls in scattered bachelor herds.

I could look out my window and see wildebeest in small herds or alone doing their thing. Taking the path of least resistance, I began observing them. Going out at daybreak every morning, I soon realized I was seeing at least some individuals, identified by slit ears, broken horns, or distinctive color and stripe patterns, every time I made the rounds. It turned out that these wildebeest were members of a resident subpopulation. Here was my chance to study wildebeest behavioral ecology by observing a sample of known individuals, essential for an in-depth behavior study. With the help of the Ngorongoro warden and game scouts, I captured, marked, and branded a dozen territorial bulls (fig. 0.5). By taking photographs of cows in the herds of females and young, I was able to identify nearly all by their stripe patterns and other distinctive marks. In addition, David Orr was pursuing his own project that entailed darting and branding a sample of some one hundred yearlings. A few of these joined herds in my study area over the course of my research.

FIGURE 0.6. Runi scanning Ngorongoro from Windy Gap, 1965.

It almost goes without saying that I couldn't devote equal time and effort to any other species. But opportunities to observe other Ngorongoro wildlife were impossible to pass up. I was often awakened at night by the rallying cries of hyenas on a kill, some involving disputes with lions. I attended a number of these encounters and often tape-recorded the ruckus. In 1964 a pack of wild dogs took up residence in the Crater for nearly a year; I *had* to observe their hunting and social behavior, including their intense competition with the 420 resident hyenas.[3,4] Before leaving in mid-1965, I had even found time to carry out a secondary, comparative study of Thomson's and Grant's gazelles.[5]

On our honeymoon, deferred a month in order to observe the wildebeest calving season, Runi and I embarked from Mombasa on a cruise ship bound for Durban in mid-January, 1965, taking my slow but trusty Land Rover pickup with us. We proceeded to observe wildebeest in the parks and game reserves of South Africa, Botswana, Zimbabwe, and Zambia on the way back to Tanzania. Returning to Ngorongoro in mid-March, we only had another three months before departure to the United States and Cornell in June. However, we had the opportunity to collect data on the annual wildebeest rut, the surround-sound and fury of which continued day and night (see chapter 10).

I intended to write my doctoral dissertation in the form of a book on the wildebeest. After publishing a summary account,[6] followed by papers on the two gazelles and the wild dog, I spent a year conducting a survey of all the different wildebeest populations. Based on the responses to a detailed questionnaire I sent to wildlife authorities of eastern and southern Africa, I wrote sixteen chapters on the wildebeest populations and subspecies in all the range states.

Problem: I had spent over a year on the survey and hadn't even begun writing up my own research. At this rate, I'd be over forty by the time I got my PhD. More important, we were planning to return to East Africa in 1968 to undertake a two-year study of the sable antelope, focusing the second year on the critically endangered giant sable of Angola. To expedite matters, I shelved the survey and wrote up only the research on wildebeest territorial behavior as my dissertation. Still, I turned forty before my PhD was awarded in 1968, some months before we arrived in Kenya to begin the sable study.

Now the truth is out: I began writing this book in the mid-1960s. Most of the 1967 survey was published separately in 2009 by the Wildlife Conservation Society, coauthored by Rod East, who had been cochair with me of the International Union of Conservation of Nature (IUCN) Antelope Specialist Group. He brought the survey up to date by adding information on the wildebeest's conservation status in the 1980s and 1990s to the end of 2005. This publication freed me to focus on the Western white-bearded wildebeest, the subspecies I know best, whose populations constitute 80 percent of all the wildebeest in Africa and define the Serengeti ecosystem. But I've included in this book parts of the 1967 survey that were left out of the 2009 publication: chapter 2 on Africa's climate and vegetation, amended where necessary to take current knowledge into account; and chapter 4, a summary of the current status of wildebeest subspecies and populations from Estes and East,[7] with a few additions of unpublished excerpts from the 1967 survey.

As it turned out, my dissertation research was only the beginning of my involvement with the wildebeest.

MORE WILDEBEEST FIELD STUDIES

1973, January–February. We returned to Ngorongoro to compare wildebeest calf survival in resident small herds and in large aggregations on calving grounds. The Munge cabin, which the National Geographic had donated to the NCAA for other researchers and VIPs to use, was

FIGURE 0.7. Munge camp where we spent two months in early 1973 observing wildebeest calving.

still standing, though neglected and considerably the worse for wear (fig. 0.7). We were accompanied by Kathryn Fuller, then a research librarian in Harvard's Museum of Comparative Zoology, later president of the World Wildlife Fund–US for a decade. Kathryn took turns with Runi looking after our son, Lindy, aged thirteen months, and assisting in our fieldwork. The presence of lions who hung out along the Munge, including a lioness with small cubs, required constant vigilance over our hors d'oeuvre-sized toddler. Indeed, the day we moved in we found a large calling card on the front porch!

A night or two later the lioness strolled into camp under a full moon and approached Kathryn's tent. As it happened, I was a witness of the event. Worried that the lions might wander in, I had gotten out of bed to look around the camp through the north window. When I saw the lioness standing still in front of Kathryn's tent, I picked up the rifle I had borrowed from NCA stores and stood watch just in case something unexpected should provoke an attack in defense of her cubs. Meanwhile, Kathryn was aware of the lions. Here's her recollection of the incident:

> I woke up in my tent on a bright night and heard and could see the form of a lioness outside my tent in the middle of camp. Her three cubs were exploring, including playing with/tripping over the tent ropes, while she kept

coughing/woofing at them. You know the sound well! In any case, I kept as still and quiet as possible, and they left without incident.

Another time, as I was returning for breakfast after checking on my wildebeest, I came on a big male walking sedately along the road that passes in front of the camp. I took the track between the road and the camp entrance, then stopped to check that the lion would continue on his way. My view of the road was screened by a patch of high weeds and grass until I was right in front of the cabin, where the vegetation had been cleared so that we could look out onto the plain. After several minutes had passed with no sign of the lion, I figured he might have decided to bed down in the high growth. I decided to flatten the patch by driving the Land Rover back and forth along its length. As row after row was "mowed," it seemed I was mistaken, but I kept on just to make sure. And sure enough, the very next to last row, the lion jumped up and walked away (fig. 0.8). The thought that he might have stayed within 30 meters of the camp while we sat outside eating breakfast and moved about was quite sobering.

Cutting the high vegetation that blocked our view from the cabin led to the other most noteworthy event of our 1973 Ngorongoro stay: a siege of bees. A previsit check of the cabin by Ngorongoro workers disclosed an active beehive between the tin roof and the softboard ceiling. They destroyed the beehive and smoked out the bees. But the bees immediately found a new home in an abandoned hammerkop nest in a tree a stone's throw from the cabin. (To see photos of typically huge hammerkop nests, use Google.) Two days after we moved in, NCA workers came to open the view, swinging pangas (machetes) with turned-up tips. For some reason, this activity set off an all-out attack by the bees, perhaps because they felt their new hive was threatened. The workers, dressed only in shorts and with a blanket covering their upper bodies, just stood there while their legs were stung over and over. Runi and I found safety in the cabin, but Kathryn was stranded in the kitchen, a bamboo structure with a thatch roof that had innumerable openings bees could enter. The workers presently got into the dump truck that had brought them and left. (We were told later that they were not seriously affected by the stings, apparently largely immune to the venom.) But we Europeans were afraid to risk becoming targets of the bees, even though our vehicle was only yards away, because many continued to dart about looking for more enemies. Our concern for Lindy was uppermost. In midafternoon we finally decided to make a getaway. We put

FIGURE o.8. The Munge pride lion observing our camp.

Lindy inside a woven laundry basket, then raced to the car, where Kathryn joined us, and we drove out onto the plain. That night we returned with Masai from the nearest manyatta who set fire to the hammerkop nest and destroyed most of the bees. That was almost the end of the ordeal, although we were attacked by survivors for the next few days. Interestingly, once you have been stung, a pheromone in the poison sac paints a target on you that stimulates other bees to sting.

1974–1978. I was employed as mammalogist at the Philadelphia Academy of Natural Sciences, during which library research and writing *The Behavior Guide to African Mammals* consumed most of my time. Coleading 1975 and 1977 academy safaris enabled me to pursue research afterward for up to two months, thanks to grants from the National Geographic Committee for Research and Exploration: first for continuation of sable and wildebeest studies. In 1977 I spent two months comparing the social and reproductive systems of topi and hartebeest with wildebeest. After spending a week in Rwanda's Akagera National Park (NP) with two Belgian researchers, Alain Monfort and Nicole Monfort-Braham, to gain insight into the topi's unusual lek-breeding system, I was scheduled to spend several weeks in Uganda's Queen Elizabeth (renamed Rwenzori) NP, where topis on the vast Ishasha Plain formed

FIGURE 0.9. The house we lived in at the Serengeti Research Institute in 1980–82.

large, mobile aggregations (both discussed in chapter 3). But the day before my flight from Nairobi to Entebbe, President Idi Amin sent out an order to round up Americans in Uganda and bring them to Kampala, the capital. That was a game changer. Instead, I spent the remaining twelve days in Serengeti NP, the Ngorongoro Conservation Area, and the Masai-Mara sampling over 2,000 antelopes to compare the survival and recruitment rates of topi, wildebeest, and hartebeest.

1979–1981. We lived for thirty months at the Serengeti Research Institute (SRI) near Seronera, headquarters of Serengeti NP (fig. 0.9). By now we had two children, Lindy, seven, and five-year-old Anna. Runi home-schooled them for those two years, so well that they slotted into the appropriate grades when we returned to the United States. These years marked the lowest point in Tanzania's economy. Tanzania had closed its border with Kenya in 1975 because Kenyan tour companies were profiting from Tanzania's wildlife and other tourist attractions, with little benefit for Tanzania. The resulting steep drop in tourism and revenue drastically limited imports of goods from Kenya. Butter, flour, eggs, cheese, milk, canned goods, batteries—even toilet paper—were all in short supply. Keeping our Toyota Land Cruiser operational was difficult because of the shortages of fuel and spare parts. The risk of getting stuck

FIGURE 0.10. Camping with the children and research assistant Marina Botje in Ngorongoro, at Tokitok springs.

or breaking down far from Seronera made going far afield slightly nerve-wracking, especially with two small children on board, as when we went camping in the western Serengeti or in Ngorongoro Crater (fig. 0.10).

To pick one of many adventures and incidents during these years, an encounter between a buffalo and our Land Cruiser was especially memorable (and expensive). Once I went on an outing accompanied by six university students staying in the SRI guest house who were interested in my topi research. Navigating a field occupied by scattered topi herds, we passed half a dozen buffalo bulls that were wallowing in a muddy pool, seeming to ignore us. A few minutes later, while we were stopped looking at the topi in front of us, a heavy blow to the rear end propelled the vehicle forward and knocked my glasses and camera to the floor, cracking the telephoto lens. At the same time, students in the cargo space behind the rear seat shouted "Buffalo, buffalo!" in high falsettos. My first thought was to get going before a second hit. Fortunately the car started immediately and we were able to leave the bull behind. The one butt of a three-quarter-ton bull stove in and crumpled the two rear doors so thoroughly that it took extensive welding to reshape them (fig. 0.11). They never again fit well enough to keep out the dust. Bad luck followed

FIGURE O.11. A wallowing bull buffalo, disturbed by our vehicle, made a profound impression.

this vehicle from the beginning. And at the end, after we had returned to the United States, a huge fever tree fell on it and made the buffalo damage seem trifling. Yet the demand for affordable 4x4s was so strong that a colleague bought it anyway; he also had reason to curse it.

This time around I was able to study several different antelopes with polygamous mating systems. The question my research was intended to answer was, what sex-linked behavioral and ecological differences caused more females than males to survive in herd-forming antelopes, resulting in adult sex ratios typically skewed in favor of females? The common resident species included topi, hartebeest, Thomson's and Grant's gazelles, and waterbuck. I focused mainly on topi and hartebeest (see chapter 3). But whenever wildebeest were within range I continued studying their behavioral ecology, as I had spent relatively little time observing the migratory Serengeti population while doing my doctoral research. In the years since, the Serengeti population had increased from a quarter million to around 1.5 million.

High points of our wildebeest observations were the three-week calving peak in February–March (see chapter 11) and the rut four months later when a quarter million bulls serviced 80 percent of the cows during a three-week peak while the migration into the woodland zone was in full swing (chapter 10). The success or failure of bull mating strategies I found particularly interesting. By observing recognizable individuals, including five prime males darted and fitted with radio collars followed day and night for up to five days, we found that certain locations in the main streams of migrating wildebeest were winning tickets in the mating sweepstakes.

1982–1995. Most of the long interval between the preceding and following extended field studies was taken up with research for and writing and rewriting my two behavior guides[8,9] and coauthoring a field guide to African wildlife.[10] However, leading camping safaris to Tanzania in 1982, 1983, 1984, 1987, 1988, 1990, and 1994 enabled me to see, smell, and hear wildebeest for two and three weeks at a time. Camping safaris I led to southern Africa in 1991, 1994, and 1995 brought renewed contact with wildebeest and other savanna wildlife of Malawi, Zambia, Botswana, Namibia, and South Africa.

The only opportunities to conduct research during this period came in 1986 and 1995. In February and March 1986, a fellowship to the University of Cape Town enabled me to conduct wildebeest research in several parks and reserves of South Africa and in Etosha National Park in Namibia, followed by three weeks in Serengeti NP and Ngorongoro Crater. The focus was on the impact of horning by wildebeest bulls on woody vegetation. I had begun collecting data on this behavior in 1979, after noticing numerous debarked and broken sapling trees following the migration of thousands of wildebeest through Seronera (see chapter 9). In June–July 1995 I spent three weeks on the Ngorongoro Crater floor collecting demographic data on all the herbivores (except hippo, elephant, and rhino). I stayed at the ranger post in the Lerai Cabin, built of stone by one of the Siedentopf brothers before World War I. It brought back memories of the good old days. This research was the prelude to the project I began in 1996.

1996–2000. From the time I started research in Ngorongoro Crater in 1963 until 1980, the wildebeest was the dominant herbivore, amounting to over half of the 20,000+ ungulates that frequented the 250 km^2 Crater floor. Between 1980 and 1986, a period when semiannual counts of Crater game were suspended, the wildebeest population declined to under 9,000. Meanwhile, the buffalo population, represented only by herds of bachelor bulls until 1970, increased to 1,500 by 1980. After counts resumed, this large ruminant continued to multiply, until it replaced the wildebeest as the dominant herbivore. Populations of the two gazelles and the eland had also undergone substantial declines (see chapter 9). By monitoring ecological conditions, population dynamics, and management practices, I hoped to identify factors that could

FIGURE 0.12. My tent, guarded by a Masai moran, in the Lemala research campsite on the northeast rim of Ngorongoro, where I camped for two months at a time from 1996 to 2002.

account for the changes. For five years I spent half the year collecting data on Ngorongoro's ungulates at different seasons: February–March, the annual birth peak of the wildebeest and other plains game; June–July, at the end of the long rains, the annual wildebeest rut and Serengeti migration; and October–November, the end of the dry season. I focused on the wildebeest, zebra, buffalo, and Grant's and Thomson's gazelles. To compare calf and yearling to adult female survival rates of Ngorongoro and Serengeti populations, I sampled both populations during the same time periods. While staying in Ngorongoro, I camped in a designated research campground on the Crater north rim and commuted to the Crater floor (fig. 0.12).

During June–July visits, I continued to follow and observe the Serengeti migration/rut. Spending nights in the pop-up tent atop my Land Rover, surrounded by thousands of croaking wildebeest, was a high point of all my years in Africa. The results of this research project were published in Estes, Atwood, and Estes.[11]

2001–2004. Although the many field studies of the wildebeest carried out in eastern and southern Africa over the past sixty years included this antelope's unusual reproductive system, the underlying reproductive physiology was largely unknown. By the late 1900s methods for tracking female reproductive cycles through laboratory analysis of hormones excreted in feces had come into use. Steven Monfort was a pioneer in this field; the laboratory he established at the Smithsonian Research and Conservation Center in Front Royal, Virginia, made it feasible to monitor the reproductive cycles of Serengeti wildebeest through regular collection of the droppings of captive cows. A grant Monfort secured from the Smithsonian Scholarly Research Program in 2001, titled "The Causation of Reproductive Synchrony in the Wildebeest," enabled the study to go forward.

But Tanzania's Wildlife Research Institute (Tawiri) refused permission to hold members of a migratory species in captivity, until we convinced the Scientific Advisory Committee that resident and migratory populations are interchangeable. Meanwhile, we had found the ideal place to keep the captives: the Grumeti Reserve, a private game reserve adjoining the northern boundary of the National Park's Western Extension (see map 5.2) leased from the Wildlife Division. Allison Moss, a graduate student of Monfort's at George Mason University, undertook the fieldwork for her doctoral dissertation. By the time a 250-hectare (ha) stretch of grassland was fenced, a campsite and field laboratory were established, and all other preparations were complete, nearly a year had passed. In November 2002 eighteen wildebeest cows were darted from among the thousands of wildebeest migrating through the Grumeti Reserve en route to the Serengeti Plain and trucked to the capture boma (corral or kraal) (fig. 0.13; see also fig. 10.9).

One objective of the research that was of special interest to me was an experiment to test the hypothesis that the calling of rutting bulls served to stimulate and synchronize estrus. That idea took root the first time I experienced the Serengeti rut in 1964. Here, at last, the opportunity to test had come. It was conducted in March–April 2003, two months before the onset of the Serengeti rut (see chapter 10).

2005–2012. By serving as resident naturalist at resorts and leading occasional safaris, I have been able to maintain contact with wildebeest for at least two months a year. Staying at Governors' Camp in the Masai Mara National Reserve during the 2005–7 dry seasons gave me the opportunity to see and film mass crossings of the Mara River. From 2008 to 2012 I enjoyed being among the hundreds of thousands of wildebeest that interrupt migration to and from the Masai Mara to

FIGURE 0.13. The migrating cows and calves captured in 2002 soon adapted to life in the Sasakwa fenced enclosure.

FIGURE 0.14. Estes family at Serengeti Wildlife Research Center at Christmas 1999, the last time we were all together in Africa at the same time and place.

spend time mowing the lush grasslands of the Grumeti and Ikorongo Reserves.

POSTSCRIPT

While my involvement in African research and conservation is pretty well over, Lyndon and Anna have both chosen to carry on research and conservation of African wildlife and ecosystems (fig. 0.14). They have earned PhDs in the Department of Environmental Sciences at the University of Virginia and are now employed at Princeton and Penn State, Lyndon as a lecturer and researcher and Anna as a postdoc. Their contributions to the survival of African ecosystems will greatly extend the contribution of their parents.

REFERENCES

1. Estes 1969.
2. Lithgow and van Lawick 2004.
3. Estes 1967b.
4. Estes and Goddard 1967.
5. Estes 1967a.
6. Estes 1966.
7. Estes and East 2009.
8. Estes 1991a.
9. Estes 1993.
10. Alden et al. 1995.
11. Estes, Atwood, and Estes 2006.

CHAPTER 1

Africa

The Real Home Where Antelopes Roam

The diversity and abundance of antelopes sets Africa apart from all the other continents. Africa has 72 to 75 different species, and Eurasia has 12; the other continents have none. The American pronghorn, adapting to similar plains habitats, looks a lot like an antelope but actually is so different that it is placed in a family of its own. Then what, exactly, is an antelope?

Good question. Antelope is the common name for all members of the family Bovidae other than cattle, sheep, or goats, plus a few tribes with no domesticated species (mountain goat and chamois, muskox and takin) (fig. 1.1). Bovids stand apart from all other ruminants—notably deer, the other major ruminant family—because males of all species and females of some have horns consisting of a bony core covered by a sheath of horn. Horn is made of keratin, the same stuff as our finger-nails. So the name "antelope" was derived through a process of elimina-tion and has no taxonomic meaning. It most likely comes from the Latin name of the Indian blackbuck, *Antilope cervicapra,* making this the only antelope deserving the name, taxonomically speaking.

Taxonomists have divided the Bovidae into different subfamilies and tribes whose genera and species share a common ancestry. But the tribes are as different from one another as cattle are from sheep and goats. This is true even of tribes placed in the same subfamily. Look at cattle, tribe Bovini, and spiral-horned antelopes, tribe Tragelaphini: they all belong to the subfamily Bovinae. The tribes in the subfamily Antilopinae are much

FIGURE 1.1 (TABLE) THE 11 TRIBES OF AFRICAN BOVIDS, COMPRISING 26
GENERA AND 75 SPECIES

Tribe	African Genera	African Species
Cephalophini: duikers	2	16
Neotragini: pygmy antelopes (dik-dik, suni, royal antelope, klipspringer, oribi)	6	13
Antilopini: gazelles, springbok, gerenuk	4	12
Reduncini: reedbuck, kob, waterbuck, lechwe	2	8
Peleini: Vaal rhebok	1	1
Hippotragini: horse antelopes (roan, sable, oryx, addax)	3	5
Alcelaphini: hartebeest, hirola, topi, blesbok, wildebeest	3	7
Aepycerotini: impala	1	1
Tragelaphini: spiral-horned antelopes (bushbuck, sitatunga, nyalu, kudu, bongo, eland)	1	9
Bovini: buffalo, cattle	1	1
Caprini: ibex, Barbary sheep	2	2
Total	26	75

SOURCE: Reprinted from Estes 1991a.
NOTE: All but 2 tribes with 3 genera and 3 species are antelopes.

more diverse. The latest classifications, based largely on DNA analyses, assign all the bovids to just these two subfamilies.[1]

In my view, lumping them this way obscures the affinities between sister tribes. Now goats and sheep are included with all the antelopes in the subfamily Antilopinae, which used to include only the gazelles (Antilopini) and the dwarf antelopes (Neotragini: dik-dik, klipspringer, etc.). Previously, goats, sheep, and goat-antelopes were cleanly separated in the subfamily Caprinae. Consequently, I'm sticking to the traditional arrangement followed in my two behavior guides[2,3] when considering the similarities and differences between the hartebeest-wildebeest tribe and other tribes of African antelopes.

The Bovidae are the latest family of ungulates to appear in the fossil record. The first known representative, *Eotragus*, dates from the Miocene, 20 million years ago (ma) (fig. 1.2). The adaptive radiation of the family tracked the global change from tropical to cooler climates and the replacement of tropical woodlands and swamps in Eurasia,

FIGURE I.2 (TABLE)   GEOLOGIC TIME SCALE OF THE CENOZOIC ERA, THE AGE OF
MAMMALS

Period	Epoch	Ma[a]	Major Events
Quarternary	Pleistocene	1.6	Ice Ages; early man; Golden Age of Mammals; maximum no. species, giant forms, followed by extinctions = 40% of species in last 20,000–10,000 years.
Tertiary	Pliocene	5	Giraffe, warthog, zebra, camel, elephant. *Homo erectus* and existing genera of bovids and carnivores appear in late Pliocene–early Pleistocene.
	Miocene	24	Early rhinos, aardvark. Ruminants become dominant as grassland habitats spread. Australopithecus in Miocene-Pliocene boundary.
	Oligocene	37	Hyrax radiation, giraffe and elephant progenitors, early carnivores, first apes, cercopithecid monkeys
	Eocene	58	Dominance of large archaic forms, replaced by late Eocene and early Oligocene by varied, mostly small ancestors of modern mammals
	Paleocene	65	Various existing and extinct orders (e.g., creodont carnivores) arise from primitive insectivore stock and diversity.
Cretaceous			Age of Reptiles

SOURCE: Reprinted from Estes 1991a.
[a]Millions of years ago.

North America, and Africa by grasslands. Most of the tribes present today evolved within the first few million years. By then, bovids were already adapted for particular biomes: gazelles and oryxes for desert and subdesert; duikers for rain forest; wildebeest/hartebeest for savannas, and so on. Goats and sheep occupied niches in the mountain ranges of the North Temperate Zone; only the Barbary sheep and two ibexes made it into sub-Saharan Africa.[4]

WHY BOVIDS RULE

There are four main reasons.

a) Being the latest herbivores to arrive on the scene. The bovid radiation reached its peak within the last several million years in the Pliocene and Pleistocene (5–2 ma; fig. 1.2). Most of the genera are still with us, though greatly reduced in numbers and geographic range.

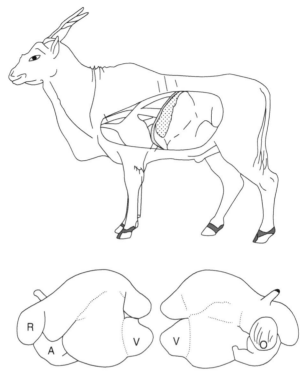

FIGURE 1.3. Ruminant stomach 5 (Hofmann 1973). Top: an eland's rumen shown in situ, overlain by the spleen (dotted organ). Bottom: an eland's stomach in left and right aspect: R, reticulum; A, abomasum; O, omasum; V, ventral blind sac. The rest is the rumen. Redrawn from Estes 1991a.

b) Advantages of the ruminant digestive system. The difference is plain to see in any pasture occupied by horses or donkeys together with cattle, sheep, or goats. The cowpats and pellets dropped by the ruminants are fine-grained, whereas the road apples of the equids contain many bits of undigested fodder. In all nonruminant herbivores, the digestion of plants occurs in the large intestine and in an appended intestinal pouch, the cecum, of which our appendix is a useless relic. Nonruminants, such as pigs, rhinos, and elephants, are hindgut fermenters, because digestion of plants occurs only after being processed in the stomach. In contrast, ruminants are foregut fermenters, as much of the nutrient processing occurs in the rumen and interconnecting stomachs (pouches) prior to entering the stomach and also afterward in the cecum, which many ruminants still retain (fig. 1.3).[5]

What I still find amazing is that vegetation eaters of whatever family or order cannot themselves digest cellulose, which forms plant cell walls, but depend on bacteria and other microorganisms in the digestive tract to perform this service. A more striking example of codependence, or mutualism, can hardly be imagined. What the ruminants did was to greatly expand the accommodation afforded these organisms in the form of the rumen, a large anteroom that is in essence a fermentation vat. Fermentation is the metabolic process of converting carbohydrates to alcohols and carbon dioxide, or organic acids under anaerobic conditions.

Some knowledge of how ruminant digestion works is important for understanding the advantages ruminants have over hindgut fermenters. The description I wrote in the introduction to the ruminants in *The Behavior Guide to African Mammals* (4) will suffice for present purposes.[2]

The rumination process is both mechanically and biochemically complex and still not fully understood. First the animal feeds until the rumen is comfortably full by gripping foliage or grass between its lower incisors and upper dentary pad, plucking, and then swallowing after briefly chewing each mouthful. Then it settles down to chew the cud, either lying or standing, grinding each bolus with rhythmic side-to-side jaw movements. To make this sideways action possible, ruminants had to lose their upper front teeth; they pluck (grazers) or snip (browsers) plants with their six lower incisors against a tough dentary pad in place of the upper incisors.

The cud consists of the coarsest plant particles, which float on top of the semiliquid rumen contents and are regurgitated a mouthful at a time through contractions of the rumen and its annex, the reticulum (the "honeycomb tripe" relished by some gourmets). As the ruminant grinds each mouthful at a steady rate, on the same or alternate sides of the mouth, enlarged salivary glands secrete a buffered solution that helps to maintain the rumen pH preferred by the microorganisms in the rumen. Chewing the cud promotes the full extraction of nutrients by increasing the surface area that is exposed to bacterial action. Some nutrients are absorbed through the rumen wall, which is lined with tongue- or finger-shaped papillae; these both vastly increase the absorptive area and provide crannies in which bacteria and protozoans multiply.

Rhythmic contractions of the rumen and reticulum keep stirring the "vat," sorting food particles according to size and specific gravity. The smallest particles sink to the bottom and from there are pumped through the reticulum into the omasum, also known as the "book organ" or "psalter" because of the leaflike plates that line it. Here the semiliquid ingesta are filtered once more before being pumped into the abomasum, the true stomach. Afterward, during passage through the intestines, the residue undergoes some final cellulose digestion in the cecum.[5]

To enable the stomach compartments to move freely, ruminants have to lie on the brisket or stand while ruminating. They rarely lie on their sides for more than a few minutes at a time. Anesthetized ruminants left lying on their sides are likely to ingest rumen contents and suffocate. The need to maintain a certain position and to keep chewing the cud may explain why ruminants do not sleep soundly as do nonruminants. This complex digestive system seemingly precludes normal sleep. Yet rumination goes together with a relaxed state (the "contented cow" syndrome), and the brain waves of ruminating animals resemble those associated with sleep in nonruminants. And until newborn ruminants begin ruminating, they sleep comfortably lying on their sides.

In addition to the more complete utilization of plant fiber in ruminant digestion, the constantly reproducing and dying rumen microorganisms that do the work provide the host with energy in the form of volatile fatty acids that they excrete as metabolic wastes, and the organisms themselves become a major source of protein as they pass through the digestive tract mixed with the rumen contents.

Ruminants possess the further important advantage of being able to recycle urea, thereby retaining and recycling inorganic nitrogen that the ruminant bacteria use to reproduce and to synthesize more protein. From this bacterial protein ruminants acquire the essential amino acids that nonruminants have to gain from their plant food. As an added bonus, recycling urea cuts down on urine excretion, helping to conserve water and contributing to the water-independence of desert-adapted ruminants.

There is one major drawback to rumination: the thorough digestion of cellulose takes time, and the more fibrous the food, the longer it takes—up to four days from ingestion to excretion. When protein content falls below 6 percent, ruminants cannot process their food fast enough to maintain their weight and condition. Hindgut fermenters consume and partially digest large quantities of low-quality forage in half the time; they can thereby manage to obtain adequate sustenance from vegetation too tough and fibrous for ruminants to process. Thus a horse can extract only two-thirds as much protein from a given quantity of herbage as a cow, but by processing twice as much in a given time, its assimilation of protein will exceed the cow's by one and a third times.

c) Habitat diversity and the ability of antelopes to specialize. Africa has by far the largest tropical landmass. It is the only continent that spans both tropics and also extends into the temperate zones to 37°N and 35°S (see map 2.1). Diversification of life zones, or biomes, began in the middle Miocene, providing new opportunities for the antelopes to exploit. The extraordinary versatility of the ruminant digestive system helps to explain how the ruminants could become so diverse, filling a greater variety of ecological niches than any other group of herbivores. The ability to structure the digestive system precisely for a given diet has enabled ruminants to partition African ecosystems into much narrower feeding niches than can nonruminants, which require a greater variety and amount of vegetation to meet their nutritional needs. For instance, the horse (Equidae) and pig (Suidae) families occupy much broader niches and have relatively few species. The bovid design for a particular ecological niche combines size, conformation, feeding apparatus (especially width of muzzle and incisor row), development and complexity of the digestive tract, dispersion pattern, and reproductive system to produce the best fit for a specific niche.

Climatic and geomorphic changes (e.g., volcanoes, rifting, plate tectonics) kept creating more habitat diversity during the Pliocene and Pleistocene, inserting new niche spaces, which antelopes proceeded to occupy.

Speciation within Africa was promoted by expansion and contraction of the Equatorial Rain Forest during wet (pluvial) and dry (interpluvial) periods of the Ice Age. During pluvial periods the rain forest stretched from coast to coast, barring interchange between northern and southern savanna and arid biomes but facilitating the dispersal of forest forms. In succeeding dry periods (Ice Ages), the rain forest was reduced to a number of isolated islands, and a drought corridor extending through eastern Africa connected the savanna and arid biomes. This explains the presence of some of the same antelopes in the Somali and South West Arid Zones, separated by the whole Miombo Woodland Zone, for instance, the oryx, hartebeest, topi/tsessebe, wildebeest, dikdik, and steenbok.

Exchanges of fauna and flora also occurred during the Ice Ages when lowered sea level created land bridges connecting Eurasia and Africa. Thus, a major faunal revolution occurred in the early Pliocene, after many Eurasian fauna crossed into Africa. Over 85 percent of the Pliocene genera were new, including nine genera of bovids, all but five unknown outside Africa. As late as the early Pleistocene, new genera continued to appear in the fossil record, due both to Asian immigrants (eight more bovid genera) and in situ evolution. The duikers, dwarf antelopes, and reedbuck tribes are the only bovids that evolved in Africa and never reached Asia.

The Sahara Desert has imposed a formidable barrier to intercontinental movement since the second half of the Pleistocene, as reflected by a much higher proportion of endemic African mammals. Eurasian species could still disperse to North Africa, but only desert-adapted forms could penetrate the Sahara. Most of the Eurasian tropical savanna fauna became extinct during the Ice Age, leaving Africa as the final refuge of the Plio-Pleistocene fauna. In the variety and abundance of large mammals—with antelopes foremost—Africa represents the Golden Age of Mammals.[6]

The areas of sub-Saharan Africa with the greatest biomass and diversity of herbivores clearly show their tropical origins. Optimum conditions occur where mean annual temperature ranges between 19° and 22°C and mean annual rainfall ranges between 750 and 1,000 millimeters (mm). These areas are associated with semiarid to moist (mesic) savanna ecosystems.[7] By far the greatest known herbivore biomass of any terrestrial ecosystem has been reached in the African grasslands.[8,9a,10,11] Even without the elephant and the hippo, the wild ungulate biomass of the best African savanna far exceeds that of recent mammals of the Great Plains, the South America savanna, the tundra, and the Eurasian steppe.[12] In an area with diverse habitats, over a dozen (up to seventeen) species may be found within a few square kilometers (sq. km) (e.g., Serengeti, Kafue, Hwange and Chobe National Parks).

Specializing for an Ecological Niche. Every kind of grazer prefers new green grass. That's only natural, as the fiber content is low and the protein content can be as high as 30 percent or more. A pasture covered with fresh green grass is every grazer's dream come true. But no grazer bites off more than it can chew, and how big a bite it can take depends on the width of its muzzle and incisor row (dental arcade). The wildebeest, with its wide muzzle and incisor row, should take bigger bites than a hartebeest or topi. The size of the bite also depends on the height and density of the grass sward. For animals the size of a topi or wildebeest, the sward has to be at least a couple of centimeters high to reward the feeding effort.

We're talking costs/benefits here. (This concept, borrowed from economics, is now pervasive in ecologists' thinking.) The first plains game to move onto a postburn flush are small selective gleaners such as Thomson's gazelle or oribi and zebra, which with their full set of choppers and flexible lips can feed as closely as gazelles. For cattle or buffalo, the stand has to be a good 10 centimeters (cm) (4 in.) high to feed efficiently. In long grass they can take the biggest bites, because with every bite their prehensile tongue sweeps out and bundles the grass. This tongue action is hard-wired in the whole tribe (Bovini) and at least some members of the sister tribe of spiral-horned antelope (Tragelaphini), most of which are browsers.

Grazers are distinguished from browsers by having high-crowned (hypsodont) cheek teeth and proportionally large stomachs. Their row of incisor teeth is also more inclined from the vertical, designed to pluck grass pressed against the dentary pad with a slight chin nod. The nearly vertical incisor row of browsers can cleanly snip foliage. Grazers need high-crowned molars because the silica content of many grasses increases the molar wear of chewing the cud. Silica (SiO_2) deposited in large quantities in the shoot system reinforces the cell walls, reducing palatability by increasing the abrasiveness of the leaves and making extraction of starch and proteins more difficult. Thus grasses defend themselves against herbivory by increasing the costs of consumption.[13,14]

Browsers are classified as concentrate selectors because their diet consists of foliage and forbs with low fiber and high protein content, making dicots more digestible with less effort than monocots. An incisor row with the two central teeth enlarged also distinguishes and enables browsers and mixed feeders to feed more selectively than grazers, whose incisors are similar in width and allow bigger but less selective bites.[15] (Fig. 1.4.)

Size is also a determining factor in the diet of grazers. Small size goes together with rapid metabolism. Larger animals have a lower metabolic rate. Because it takes more time to ferment and break down grass, most pure grazers are of medium to large size. The smallest grazing antelope is the 14 kilogram (kg) (30 lb.) oribi. So it is no accident that the bulk feeders are medium to large, as it can take several days to extract the nutrients from fibrous grasses.

The number of bites a minute is another measure of feeding efficiency. The ecological concept of optimal foraging applies here. In theory, natural selection should lead to the most efficient feeding behavior a species can achieve. Thus a grazing animal should feed at the fastest possible rate in the pasture of its choice. This rate is species-specific, dependent on size, feeding apparatus, and selectivity of grass stage. The quickest eaters take up to thirty thousand bites in a day.[16]

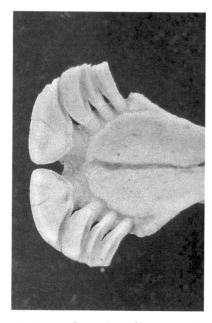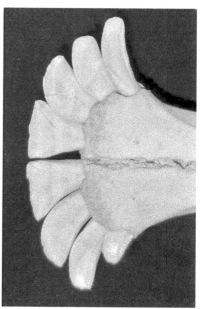

FIGURE 1.4. Comparison of browser and grazer incisor arcs. Note the enlarged central incisors of the browser compared to the wide, evenly developed incisor row of the wildebeest. The two incisors on either side are modified canines common to all bovids—the left one still erupting. Photos provided by Harvard MCZ Mammal Department.

According to Ledger,[17]

Tropical wild animals appear to be physiologically incapable of depositing large amounts of fat, in the carcass, thorax, or viscera, even when ample food is available. Conversely, there is so far no evidence that their body composition is materially affected when food is less plentiful. The evidence suggests that game animals may be physiologically geared to a near-maintenance level of production and that once these requirements have been met, their appetite is satisfied and they cease to feed. From a meat production point of view this represents a degree of inefficiency because such animals are then incapable of making use of an abundance of food. From a soil conservation standpoint such a limitation may be an advantage as it means that the vegetative cover is then easier to maintain. (138)

Putting all this together, we see that adaptation for a particular ecological niche involves a wide range of different factors. But when all is said and done, size and digestive system stand out as the major determinants of niche separation.[18,19]

d) The dominance of the grazers. Three tribes of grazing ruminants dominate Africa's savannas.

Reduncini (reedbuck, kob, waterbuck) occupy valley and floodplain grasslands within a short distance of water.

Hippotragini include the desert-adapted addax and oryxes, as well as the water-dependent sable and roan, which inhabit the well-watered savanna woodlands.

Alcelaphini exploit savanna ecosystems with extended wet and dry seasons. They disperse into dry savanna during the rainy part of the year and concentrate on green pastures near water in the dry season.

The Bovini are the fourth tribe that falls into this category, but the Cape buffalo *(Syncerus caffer)* is the only native African bovine. Bovini evolved in Eurasia. The Sahara was an impenetrable barrier to these water-dependent grazers. Ironically, domestic cattle now far outnumber all Africa's wild bovids but only because of human intervention.

The members of these tribes include the most advanced ruminants. They have the most developed digestive systems, capable of subsisting on dry hay with as little as 6 percent protein content. They are of medium to large size, built to travel long distances in search of green pastures. They prefer open over wooded habitats and flee rather than hide from predators. They include the most gregarious African bovids with the most conspicuous coloration and markings.[20] A few species (wildebeest, topi, blesbok, kob, lechwe, buffalo) form large aggregations and dominate whole ecosystems. These are the keystone species.

Bovid ancestors were small, solitary dwellers in forests and other closed habitats where they relied on concealment to avoid predators. Duikers like the blue duiker and red duiker and dwarf antelopes like the dik-dik and steenbok are most like the early bovids.

To leave cover and exploit the abundant grasslands required a whole suite of profound changes:

Size and conformation

From hiding to avoid predation to reliance on flight in the open

From browsing to grazing

From solitary and pair-forming social systems to herding and polygyny

From cryptic to conspicuous coloration

Development of visual displays as a major avenue of communication

The Division between Bovids of Closed and Open Habitats

Many of the morphological and behavioral adaptations to closed and open habitats are opposite. Thus antelopes of closed habitats that hide to avoid predation have hindquarters that are more developed than forequarters and cryptic or disruptive coloration and markings (duikers, dik-dik, bushbuck, kudu). This conformation enables quick starts, sharp turns, and high bounds when an animal is flushed from its hiding place. Antelopes of open habitats have long, evenly developed limbs (gazelles, oryx) or higher, more developed forequarters (hartebeest tribe) adapted for long-distance travel and the speed and endurance necessary to outrun predators.

In an analysis of the correlations between conformation (phenotype), socioecology, and mating system of African bovids,[21] I sorted the bovids into two distinct ecotypes, one adapted for life in closed habitats and the other adapted for life in the open. It appears that this dichotomy applies to the whole family.[2,20]

Group I, comprising bovids that live in closed habitats, have cryptic and disruptive coloration, and possess the conformation just described, is associated with a concealment strategy. The duikers and dwarf antelopes are solitary or live in monogamous pairs and have minimal sexual dimorphism and approximately equal adult sex ratios. Two reedbucks fall into this category but are of medium size (up to 80 kg [175 lb.]) with well-developed horns in males only. Most members of the bushbuck/kudu tribe (Tragelaphini) also belong in Group I but are larger, polygynous, and very sexually dimorphic, with skewed adult sex ratios.

In Group II, comprising bovids that live in open habitats, antipredator strategies depend on the ability to outrun predators in open flight, or on refuges (cliffs for goats, swamps for lechwes and sitatunga), or on large size and/or group defense (buffalo, other Bovini). All the member species are gregarious, have polygynous mating systems with adult sex ratios favoring females, and have at least some degree of sexual dimorphism (males larger, with more developed horns). The most conspicuous and distinctive coloration is seen in the most gregarious species that live in the most open habitats. Hippotragini, Alcelaphini, and most Reduncini, the gazelle tribe, goats and sheep, and bovines (buffalo, etc.) belong in this category.

How so? When there is no need to hide, selection for visual communication causes conspicuous coloration to evolve.[21,20] The need to be as conspicuous as possible to conspecifics and different from other associated bovids can account for the color schemes that distinguish bovids of open habitats: dark or light coats that contrast with the natural background, black and white markings that contrast with coat color, and reverse countershading (darker below, lighter above, as in the wildebeest). Furthermore, coloration is only one of a suite of visual recognition characters, including size, conformation, locomotion, horns, and other appendages, so redundant that associated species can be recognized by silhouette alone. Such redundancy of recognition characters is in itself evidence of the premium placed on development of a distinctive appearance that promotes communication and mating with one's own species while selecting against attraction to the traits of other species.[22,23] The contrasting coloration and conformation of topi and hartebeest subspecies are illustrated in chapter 3 as cases in point.

That virtually all bovids that live away from cover are social while virtually all solitary species depend on concealment suggests that gregariousness is an essential adaptation for life in the open. One might go further and postulate that herd formation was a prerequisite to abandoning concealment. Antelopes band together in the open for mutual security: the herd takes the place of cover for the individual. Take away their cover, and even Group I antelopes like reedbuck and oribi join in temporary herds. In short, the mutual security offered by a group may be the foundation of sociability.

Leaving aside sick or injured animals, the only individuals of gregarious species seen alone are mothers keeping watch over concealed young calves and males guarding territories. The willingness of males to stay alone places them at greater risk but is a prerequisite for reproduction in territorial mating systems.

To become sociable it was necessary for females to forgo territoriality. In fact, male territoriality is the rule in all but nine African antelopes, all of them in the bushbuck-kudu tribe (Tragelaphini). This tribe is allied with the buffalo-cattle tribe in the subfamily Bovinae, whose lineage has been separate from the Antilopinae subfamily (as currently classified) since the Miocene.[24] Instead of territoriality, a male dominance hierarchy, based on prolonged physical development, determines mating success.

CONSEQUENCES OF THE MALE ARMS RACE

Competition between males to monopolize breeding opportunities with females in herds selects for development of characteristics such as size, weapons, and ornamentation that make for dominance. The larger the herds of females, the more intense the male competition. It is the classical peer competition Darwin described.[25] When the reproductive unit is a mated pair, competition is minimal, and so are physical differences between the sexes. In fact, female duikers are slightly larger than males, and both sexes have horns. In solitary and monogamous systems, grown offspring of both sexes perforce leave home. Facing equal risks, mortality rates are similar and the adult sex ratio remains more or less equal. In herd-forming species, the intolerance of breeding males toward perceived rivals forces male offspring to leave their natal herd and home range, whereas females can remain at home. Consequently, males suffer higher mortality and the adult sex ratio is skewed toward females, supporting the polygynous (multifemale) mating system.

Among the male secondary sex characters that evolve through peer competition, development of large and elaborate horns makes for the most obvious gender difference, especially in species with hornless females. In the impala, for instance, color and markings are the same in both sexes. But in the Reduncini, in which females are also hornless, not only the horns but also male coloration and markings of the most gregarious species (white-eared kob, lechwe) enhance gender differences (sexual dimorphism). In species with horned females, male horns are typically far more developed than in females, as seen in the gazelles, sheep, and goats (fig. 1.5).

Color dimorphism is pronounced in the sable antelope, alone of all the Hippotragini: mature males are jet black with white markings, while females stay brown. In order to stand out in the crowd to maximum effect, males need to keep growing for up to three years longer than females. Sable cows conceive at two years, while males finally mature at five years. But the most extreme maturation difference is seen in the nonterritorial, male-dominance mating system. Thus eland cows also conceive in their third year, whereas bulls mature at eight but have to keep growing in order to compete with their seniors, who slowly but surely continue to add bulk.

HOW TO EXPLAIN MINIMAL SEXUAL DIMORPHISM IN WILDEBEEST AND ORYX TRIBES

Except for the sable, antelopes in the Hippotragini and Alcelaphini display surprisingly little sexual dimorphism (SD). As they too have territorial, polygynous mating systems, males would be expected to respond

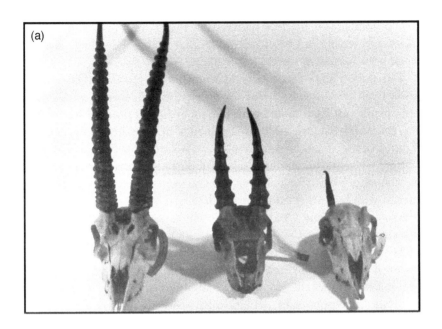

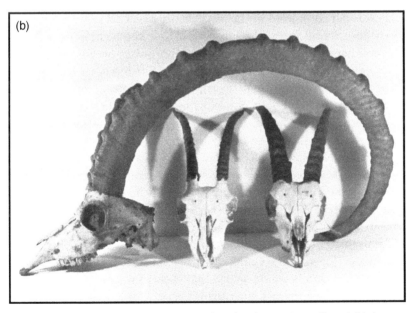

FIGURE 1.5. Gender contrasts in the horns of (a) the Thomson's gazelle and (b) the ibex. Female Thomson's gazelle horns do not mimic horns of even the youngest males (yearling in the middle). That could be why tommy females are losing their horns. Contrast in male and female ibex horns is the outcome of extreme male peer competition seen in sheep and goats, with mating systems based on a male dominance hierarchy. Nevertheless, the photo shows that female horns mimic the early growth stages of male horns. Photos reprinted from Estes 1991b.

to peer competition by clearly advertising their sex. Yet the similarity of males and female can be close enough that sexing individuals is challenging for human observers (fig. 1.6).

The notion that females might mimic male secondary characters arose from the difficulty I had sexing passing wildebeest: cows had a faux penile tuft that looked a lot like the male's real one. I wondered, was this annoying little addition designed to deceive other wildebeest as it deceived me? Years later, an analysis of African antelope social organization[21] showed that minimal sexual dimorphism in gregarious species went together with the habit of forming mixed herds, particularly in species with migratory and nomadic habits. This correlation suggested that minimal SD could enable males to mingle with minimal aggression. It seemed that some force countered selection of conspicuous male secondary characters, leading to reduced SD and intolerance between males. It was this male intolerance that led to the expulsion of adolescent males from female society, as I had observed countless times.

Peer competition is by definition between equals. But competition between different age classes is by nature unequal. Aggression by an adult against a younger male is a form of *despotic competition*.[27] This form was not considered in Darwin's theory of sexual selection and has rarely been considered as the antithesis of peer competition since. Yet it can counteract peer competition by suppressing or postponing the development of male secondary characters.[2,27,28] In his seminal treatise on sexual selection, Darwin actually pointed out an example of despotic competition that he could not explain: the horns of yearling elands were more developed than the horns of yearling male kudus, whose horns are among the largest of any bovid.

The plausible explanation is that male kudus cannot afford to advertise their sex by developing this most obvious male secondary character before they are grown up enough to survive eviction from the maternal herd. Both sexes of the eland bear horns. Females and young males look enough alike that males do not stand out until their third year or even later.

The same holds true of other bovids. As horns of both sexes are equally developed in the young of horned females, males are tolerated until they develop beyond female limits. Being thus disguised, they can begin horn development sooner than species with hornless females—in response to peer competition. That would explain why horn development begins earlier in all bovids with horned females, typically within the first two months, compared to six months or so in offspring of hornless females.[26]

FIGURE 1.6.
Comparison of
male (a) and
female (b)
wildebeest. c)
Male and female
gemsbok. Minimal
sexual dimorphism
in oryx/gemsbok is
explainable by
females tracking
male secondary
sexual characters
to the adult stage.
SD in wildebeest is
less complete, but
note the male-like
penile tuft typical
of females of the
Serengeti
population.

This train of evidence led to the hypothesis that females evolved horns to buffer male offspring from despotic competition by mimicking this iconic male secondary character. How else could they disguise this permanent appendage? Fossil remains show that females of the earliest bovids were hornless and that horns evolved in females of the different tribes independently.[29] Females in thirty-one of the seventy-two African antelopes remain hornless.

The conventional view of female horns is that they evolved as weapons of self-defense and offspring defense against predators.[30,31] Certainly females use them as weapons but so seldom and ineffectually against predators that their advantage over hornless females is hard to see or quantify.[26,32] Prolonging the stay of male offspring, on the other hand, should be a strong source of both natural and sexual selection. It enables sons to stay in the maternal herd until the benefits of leaving outweigh the costs of staying while at the same time getting a head start (!) on horn development. In this way mothers also promote their sons' reproductive success, which is potentially much greater than that of daughters, who produce only one offspring a year.

If the horn mimicry hypothesis sounds far-fetched, consider the fact that male mimicry is not limited to horns. Females also replicate beards, manes, dewlaps, and coloration to lesser or greater degree. And don't forget the wildebeest cow's penile tuft (see chapter 4).* But horn mimicry stands out as most essential because males cannot compete without them, and their size is important in reproductive competition.

I find it fascinating that male sexual competition in the bovids can have opposite effects depending on whether it is between peers or between older and younger males. It is despotic competition that checks and slows the development of horns and other male secondary characters that arouse male intolerance. As long as males and females look and act alike, despotic competition is minimal. In species with pronounced SD, females stop mimicking males at the stage where the cost-benefits favor their joining all-male herds wherein they can openly engage in peer competition.

The minimal SD exemplified by wildebeest and oryx is the outcome of females tracking male secondary characters to the adult stage. As already noted, reduced SD correlates with social groups that include

---

* Apparently the Western white-bearded race is the only subspecies that has developed the penile tuft. See chapter 4.

adults of both sexes. In the desert-dwelling oryx, the chances of evicted males finding a bachelor herd are so slight that natural selection favored remaining permanently in the company of females. The resemblance of the sexes is so close that the alpha male treats both sexes like females.

How far females track male characters depends on the species' social and ecological circumstances. Horn mimicry extends no further than early adolescence in the Thomson's gazelle and in various sheep and goats, in which females have short, thin spikes for horns, all the way to maturity in the Alcelaphini and Hippotragini, whose horns, though never as robust as in males, are of similar shape and size.

For an in-depth review of the male mimicry hypothesis, see Estes.[26] In a later paper,[20] I propose that the evolution of conspicuous coloration and markings in bovids with gregarious, polygynous social systems was promoted by continuing female mimicry. That could happen because of peer competition to develop distinctively male secondary characters. If females proceed to track these characters, then males have to keep trying to be different. When females copy these features to the adult stage, you end up with a gaudy animal like the gemsbok, the most conspicuous and least dimorphic antelope.

REFERENCES

1. Gentry 1990.
2. Estes 1991a.
3. Estes 1999b.
4. Kingdon 1982.
5. Hofmann 1973.
6. Huxley 1965.
7. Thackeray 1995.
8. Bourlière and Verschuren 1960.
9. Talbot and Talbot 1963.
10. Lamprey 1964.
11. Field and Laws 1970.
12. Bourlière 1961.
13. Kaufman et al. 1958.
14. Hunt et al. 2008.
15. Janis and Ehrhardt 1988.
16. Murray and Illius 1996.
17. Ledger 1964a.
18. Bell 1970.
19. Jarman 1974.
20. Estes 2000.
21. Estes 1974.

22. Dobzhansky 1937.
23. Mayr 1963.
24. Kingdon 1997.
25. Darwin 1871.
26. Estes 1991b.
27. Schein 1975.
28. Wilson 1975.
29. Janis 1982.
30. Packer 1983.
31. Stankowich and Caro 2009.
32. Walther 1966.
33. Brooks 1961.
34. Estes 1967a.

African Savannas

Understanding the Tropical Climate,
Vegetation, and the Gnu's Ecological Niche

This chapter is intended as a primer on Africa's climate and vegetation as an aid to understanding the kinds of habitat that constitute the wildebeest's ecological niche. Inhabitants of the north temperate zone often have little understanding of tropical climates beyond knowing it's warm there. The position of the sun directly overhead at the equator and between the Tropics of Cancer and Capricorn at 23.5°N and S is the reason this zone is the warmest part of the globe (see map. 2.1, below). Due to the tilt of the earth's axis, the sun appears to move between the two tropics, reaching its northern limit on June 21, the summer solstice, and its southern limit on December 21, the winter solstice. Africa is the only continent that spans both tropics. Encompassing some 70 percent of the whole landmass, the tropical region is so vast that the bordering temperate zones are crowded at the north and south ends.

The earth is warmest at the latitude where the sun is vertically overhead. This heat equator changes as different parts of the earth are exposed to the high sun. Warm air rises, and cool air sinks. Warming of the earth beneath the sun makes the overlying air rise, creating a zone of low pressure at the surface. Cooler air flows into the void from either side; the greater the difference in pressure, the stronger the flow. Their convergence creates the trade winds, which due to the earth's rotation are deflected clockwise in the northern hemisphere and counterclockwise in the southern hemisphere. Northeast trade winds prevail while the sun is north of the equator, and southeast trades prevail while it is

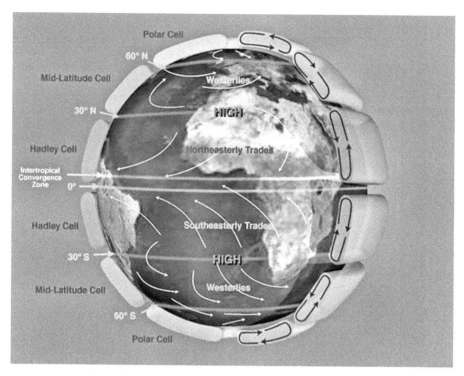

FIGURE 2.1 Atmospheric circulation over Africa. Intertropical convergence zone (ITCZ), created by convergence of northeast and southeast trade winds at the heat equator, follows the sun north to the Tropic of Cancer during northern summer and south as far as the Tropic of Capricorn during northern winter. Hadley cells represent rising of heated air at the ITCZ to the upper atmosphere, then flowing north and south to descend at the tropics and flow back to the ITCZ. *Wikipedia Commons* http://sealevel.jpl.nasa.gov/overview/climate-climatic.

south of the equator. The zone where the winds converge is known as the intertropical convergence zone (ITCZ) (fig. 2.1). There moisture sucked from the earth condenses into clouds as cooling shrinks the volume of rising air, producing rain when condensing water vapor turns to heavier than air droplets.

The air in the ITCZ rises some 9 mi. (14 km) to the upper limit of earth's atmosphere (the troposphere). Then it flows horizontally toward the poles, descending and finally coming back to earth at the borders between the tropics and temperate zones. Deserts occur there on all continents, because the increasing capacity of descending air to hold moisture prevents condensation and rain. Higher pressure in the arid zones where the sinking air meets the earth causes a return airflow, completing the circuit. This circulating system is known as a Hadley cell. These cells occur in both tropics. Two other primary circulation cells, created by air moving between large high- and low-pressure regions, encircle the globe: Ferrel cells, in the temperate zone of westerlies, and the Polar cell, with prevailing northeast winds. The location of these belts and the prevailing wind direction are determined by solar radiation and the rotation of the earth.

The atmospheric circulation that produces Hadley cells through convection and the alternation of the trade winds determine the regional and seasonal patterns of tropical temperature and rainfall. A third driving force is the El Niño–southern oscillation of the Pacific Ocean. Though more remote in origin, it strongly influences year to year rainfall and temperature patterns in tropical Africa.[1]

THE SAVANNA BIOME

Between the equatorial rain forests with only brief dry seasons and the deserts with no regular rainy season are the savannas, where grass grows in association and competition with trees. It is the presence of a continuous grass layer that distinguishes even dense woodland from forest. Grass and other undergrowth grow wherever sunlight reaches the soil. The canopies of savanna trees are open enough to let sunlight penetrate. Savanna occupies some 37 percent of the African continent, compared to 39 percent desert, 18 percent forest, and 5 percent open plains (treeless grassland).[2,3]

Savannas can range from very sparse tree cover to dense woodland and bush, and the relative proportions of grass and trees are determined by competition between the two life-forms for water and nutrient resources. Underlying abiotic factors such as rainfall, soil depth and composition, topography, and disturbances such as fire, herbivory, and extreme weather events form a complex web of interacting factors that influence this competitive interaction. The balance of forces determines savanna structure at a given location.

The vegetation map (map 2.1) reveals a certain symmetry in the arrangement of the vegetation zones on either side of the equator. In both directions, equatorial rain forest yields to forest-savanna mosaic to moist savanna and arid savanna to subdesert and finally to desert approaching the Tropics of Cancer and Capricorn. This symmetry is a reflection of similar climatic regimes. There is also dissymmetry, as is

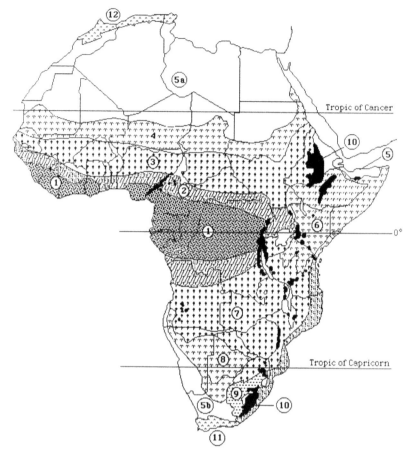

MAP 2.1. Africa's main vegetation types. 1) West African and Congo Basin rain forest.
2) Forest-Savanna Transition and Mosaic. 2a) Coastal Forest-Savanna Mosaic.
3) Northern Savanna (including Guinea Savanna and part of Sudanese Semi-arid
Transition Zone). 4) The Sahel (arid savanna and bushland). 5) Desert and semidesert
shrubland. 5a) Sahara Desert. 5b) Namib Desert (including karoo semidesert).
6) Somali-Masai Arid Zone (dry savanna). 7) Southern Savanna (including wetter
miombo woodland and drier acacia savanna). Southwest Arid Zone (including Kalahari
and karoo dwarf shrubland). 9) Highveld temperate grassland and karoo grassy
shrubland. 10) Afromontane vegetation (from montane forest to montane grassland,
bamboo, heath, and alpine zones). 11) Cape and karoo winter-rainfall shrubland
(Cape Macchia). 12) Mediterreanean winter-rainfall vegetation. Reprinted from
Estes 1991a. Based on White 1983.

expected considering the huge landmass of northern Africa and the narrowing of the subcontinent. The way the vegetation zones bend north and south in northeastern Africa reflects not only the much greater landmass but also the elevated inland plateau that comprises most of eastern and southern Africa.

A defining feature of a tropical climate is seasonal rainfall, with precipitation limited to one or two definite rainy seasons, followed by a lengthy dry season. In the temperate zone, rainfall tends to be distributed more evenly and the weather is far more variable due to traveling high- and low-pressure systems, or fronts. Traveling lows and highs rarely develop in the tropics, but the strength of the subtropical highs is directly affected by the traveling low- and high-pressure fronts of higher latitudes. After all, there are no impermeable barriers to the atmosphere. What is freer than air? The far greater landmass of northern Africa causes it to heat up more than the subcontinent, so that the ITCZ extends to about 20°N latitude in July, compared to about 15°S latitude in January.[4,5] Although high pressure also forms over the arid Kalahari during the southern winter, it weakens to a greater or lesser extent in summer, permitting rainfall to develop. Only the Namib Desert, kept rainless by the cold air over the Benguela Current, qualifies as true desert. And during an El Niño event that brought incredible rains to Namibia in 2011, it actually rained at the coast when I was there in January.

At the equator, the sun passes overhead twice a year during the equinoxes, resulting in two rainy seasons. In the low country of West Africa within about 4° of the equator, the two rainy seasons are so protracted that they overlap and the intervening dry seasons occur only as months of reduced rainfall. The lowland rain forest coincides with this rainfall regime in West and Central Africa. In equatorial East Africa, the elevated plateau and topographic features such as high mountain ranges, the Rift Valley and Lake Victoria, make for a more varied and generally much drier climate, so that the two seasons are well marked. February, coming between the short and long rainy seasons, is the driest month in Kenya and Uganda. Proceeding from Rwanda eastward and from southern Tanzania to Somalia, the country becomes ever more desiccated, with subdesert scrub extending to within a few miles of a comparatively lush coastal strip on the Indian Ocean. Yet only a few degrees south of the equator, in southern Tanzania, the two rainy seasons give way to one long one, followed by a long dry season. This is the regime that prevails over the rest of the subcontinent, except for the very tip of the Cape of Good Hope, which has a Mediterranean winter rainfall regime.

Tropical weather may seem regular compared to the synoptic weather of the temperate zone, but it is very complex and subject to great variation.[4,5] The differences in the amount of precipitation from one rainy season to the next are often profound, as later examples illustrate. It is axiomatic that the more arid the climate, the more variable and erratic the rainfall. As already noted, the most profound outside influence on African rainfall results from warming or cooling of surface waters in the tropical Pacific Ocean. Every five years on average, the eastern Pacific becomes warmer or cooler than usual. Warming of the overlying air mass, called the southern oscillation, increases the surface pressure, resulting in an El Niño event. For instance, the El Niño of 2010 brought devastating floods to East Africa. The heavy rains that inundated northern Kenya followed hard on the heels of a terrible drought connected with a La Niña event that occurred when the ocean temperature in the eastern Pacific became cooler than usual. (See Google ENSO on NOAA website for details.)

This may be enough information about Africa's tropical climate for present purposes. I need only add here that local precipitation may occur outside of the main rainy season wherever there is sufficient moisture in the air and convection currents to convey it to the cold upper atmosphere. Thus mountains receive enough precipitation from rain and fog to support lush cloud forest in the midst of desert (as in northern Kenya) and parched plains in the vicinity of lakes, or the ocean may receive heavy downpours from thunderclouds that form due to random thermal convection over heated ground. These are the same kinds of storms so familiar in the temperate zone during summer months. Arid environments usually receive their precipitation in this form and rarely experience generalized rain.

WHAT MAKES A SAVANNA?

The length of the dry season and how much rain falls in the wet season(s) determine the type of savanna. Proceeding north, south, and east from the West African rain forest, average annual rainfall steadily decreases and the length of the dry season increases, from nil to as much as ten or eleven months in subdesert and a full twelve months in most of the Sahara. The vegetation shows a corresponding progression: from tall forest with a closed canopy rich in epiphytes but with almost no ground cover; to a forest-savanna mosaic, where the forest is divided into blocks by openings dominated by high grass and bushes; to progressively more open, wooded savanna with smaller trees, which grade from evergreen to broad-leaved deciduous to small-leaved thorny species; to steppelike landscapes too arid to support trees except along drainage lines, where more and more of the ground is laid bare; and finally to true desert, where the soil may be almost entirely bare save for scattered clumps of low shrubs and bushes.

Like the trees, the grasses undergo a progressive reduction in height, density, width of leaf, and basal diameter with increasing aridity. There is a gradual thinning and stunting of the vegetation, a general impoverishment of the environment, but with no sharp boundaries between vegetation zones. It is a continuum in which types intergrade, so that defining and demarcating types is a more or less subjective process. White's (1983) vegetation map of Africa,[6] published by UNESCO, is still one of the most authoritative.

OUTLINE OF THE VEGETATION ZONES BETWEEN THE EQUATOR AND THE DESERTS

Broad-leaved, deciduous savanna woodland (or wooded savanna), with trees up to 60–70 ft. (18–21 m) high, rainfall of 36–45 in. (914–1,143 mm), and a dry season of four to six months. This is the most homogeneous and extensive major African vegetation zone, occupying the interior plateau at 3,000–5,000 ft. (915–1,525 m) all the way from southern Tanzania and most of Mozambique to central Angola, a belt of open woodland 1,000 mi. (1,609 km) from north to south by 800 mi. (1,287 km) east to west. Its northern edge corresponds closely to the latitude at which a single rainy season begins, for the dominant plants of this type are adapted to one-season summer rainfall with a long, severe dry season.[6] It is known as *Brachystegia-Julbernardia (Isoberlinia)* woodland for the two dominant genera of (leguminous) trees or as *miombo* (Kiswahili for *Brachystegia* trees), the name for this biome in southern Tanzania.

The so-called Northern or Guinea Savanna is very similar to *miombo* woodland and also very extensive, stretching over 3,000 mi. (4,828 km) from west to east but only 250–300 mi. (402–483 km) from south to north to merge with the Sudanese dry savanna. Tall, coarse grasses predominate, especially *Hyparrhenia* and *Andropogon* species; these carry hot fires through the woodland every dry season, so that only fire-resistant trees can survive in this environment. In both of these wooded savannas, the ground retains enough moisture to produce new grass and forbs on seasonally flooded soils, and deep-rooted trees flower and come into leaf before the rainy season begins—in the so-called *miombo* spring. That doesn't happen in more arid savannas.

Semiarid (mixed) savanna, with trees up to 45 ft. (14 m), rainfall of 20–36 in. (508–914 mm) and a dry season of six to eight and a half months. The transition from *miombo* to mixed savanna is seen in more open savanna with smaller trees and in the gradual decrease of broad-leaved deciduous trees and an increase of acacias and other thorny trees and bushes. But the proportion of trees to grass is infinitely variable, ranging from a continuous light canopy to parkland with scattered clumps of trees to grassland with scattered trees to treeless plains. The distinction between wooded savanna (or savanna woodland) and tree savanna is therefore arbitrary, as there is no definite break in the progression from woodland to open grassland. Even in wooded savanna, the woody cover is seldom over 50 percent; the trees stand in grass instead of forming a canopy over it and are generally small, deciduous, and spindly.[7]

Semiarid savanna is more extensive in the north than in the south. Nearly spanning the widest part of the continent as a band 150–350 mi. (241–563 km) from south to north, it is called the Sudanese Semi-arid Transition Zone.[8] In southern Africa this zone is referred to as bush-veld, *veld* being the Dutch/Afrikaans term for savanna. It occupies a broad but irregular region trending diagonally from the Natal coastal hinterland to southern Angola, a distance of around 1,500 mi. (2,414 km). The habitat diversity of this zone is revealed by such adjectives as mixed, transitional, or undifferentiated, relatively dry types. Indeed it is shown differently on every new map. The quite densely wooded lowveld of the eastern Transvaal and Natal is included in this zone, as well as sparse tree and bush savanna in the increasingly arid lands of western Botswana and Namibia. Mopaneveld, named for a broad-leaved tree with bilobed leaves *(Colophospermum mopane)*, is also included in this zone.

Arid savanna, with 4–20 in. (102–508 mm) of rainfall and a dry season of eight and a half to ten months. Woody growth ranges from treeless open plains to tree savanna, parkland, open wooded savanna, and even dense thickets. The woody vegetation is predominantly thorny, dominated by various species of *Acacia.* But in the better-watered areas, such broad-leaved deciduous species as *Commiphora* and *Terminalia* rival acacias for dominance. The most extensive arid savanna is the Sahel, which lies between the **Sudanese Semi-arid Transition Zone** and the Sahara subdesert. In East Africa, the **Somali-Masai Arid Zone** includes the classic acacia savannas and plains teeming with big game. The third extensive region of arid savanna grows on the Kalahari sands of Botswana, northern South Africa, and Namibia. It is an ecotype that resembles and rivals the acacia savannas of Kenya and Tanzania.

Subdesert, with rainfall of 4–15 in. (102–381 mm) and a ten- to twelve-month dry season, merges imperceptibly into arid savanna on one side and desert on the other. It occurs mainly at low altitudes, where higher temperatures, evaporation, and runoff reduce rainfall effectiveness. Trees are confined to watercourses, and the vegetation consists mainly of drought-resistant bushes, succulents, and tubers. Most grasses are annuals, with *Aristida* spp. predominating. Along with an extraordinary variety of ephemerals, they cover the ground for a few weeks or months after a rare rainfall. Aside from forming a zone below the Sahara, subdesert covers most of northern Kenya, Uganda, and Somalia. In these tropical subdeserts, *Acacia* and *Commiphora* spp. manage to grow as low shrubs. In South Africa, this zone is called the **karoo** for the variety of low woody karooid shrubs it contains, and lies in the south temperate zone, with winter frosts. Karoo adjoins the Kalahari **sandveld** on the north and in Namibia lies between the Kalahari and Namib Deserts.

Although the general trend from moist to dry savanna is toward more open country with shorter grass, all gradations from woodland to open grassland occur in every type, depending on the nature of the soil, which is determined by slope and drainage as well as the character of the underlying soil-parent material. On sloping ground such as hillsides, a sequence of different soil profiles occurs, descending the slope. The soil grades from coarse and well drained on the upper slopes to fine-grained clay soil with impeded drainage at the bottom. Soils of similar structure are associated with very similar plant communities. This repeating sequence is called a catena. Woodland and short grassland dominate on the upper slopes, becoming more open as the gradient levels out, yielding to taller grassland on the flat and finally to marshland or swamp in undrained depressions (see fig 9.2).

MODIFICATION OF SAVANNA BY HUMANS

The equatorial rain forest of West Africa, most of which lies in the Congo Basin, delineates the region in which there is no protracted dry season. Certain areas, as on the Liberian and Cameroun coasts, receive over 150 in. (3,810 mm) a year, but an average of 68 in. (1,727 mm) suffices to maintain an unbroken forest canopy—provided the rain is evenly distributed. Past a certain point, then, rainfall distribution is more important than quantity in determining the character of the vegetation. Even with a rainfall of up to 60 in. (1,524 mm) a year, three dry months is enough to inhibit rain forest and to create the conditions for savanna, although of course other factors, including higher temperatures and less cloud cover in the tropics, are involved.

Once the forest canopy is opened up enough to allow sunlight to reach the floor—whether due to drought, cutting, or burning—grass and other undergrowth can establish. In areas with no real dry season, the forest quickly regenerates. But if there is a dry season of sufficient rigor to stop grass from growing, then it is almost certain to be burned every year. In this manner fire-tender trees are destroyed, the canopy is further opened, the proportion of grass increases, and the balance shifts further toward savanna.

Man-made fire has been such a universal factor in savanna ecosystems for so long that it is often impossible to determine how much of the African savanna is the result of human impact. Until recently, it was generally accepted by ecologists that, in the absence of expert supervision, human influence in the form of burning, cutting, stock grazing, and cultivation has almost invariably been in the direction of downgrading the environment from complex to simple communities, turning forest into scrub or savanna, moist savanna into arid savanna, and arid savanna into subdesert scrub. In other words, human influence has been considered to degrade habitats. Some ecologists have gone so far as to argue that practically all African grasslands have been created and maintained by humans, primarily through fire.[9,10]

Others divide savannas into two categories: those maintained by fire in higher rainfall regions that would otherwise revert to woodland; and those in more arid regions where tree savanna and open grassland occur naturally.[10,11] While periodic burning may be necessary to maintain grassland where woody vegetation is the natural climax, there is good evidence that climate and soil conditions interact to produce savanna and pure grassland over wide areas absent human agency. Fires started by lightning are common in southern Africa but rare near the equator. I was surprised to learn that no fire started by lightning has ever been recorded in the Serengeti,[12] where almost all fires are set by humans, including as management tools in parks and reserves (see chapters 9 and

12). Although lightning strikes certainly occur during thunderstorms, rain follows closely and quickly douses the flames.

Recent studies of savanna ecosystems from a pastoralist perspective challenge the assumption that humans and their livestock degrade rangelands. In her recent book, *Savannas of Our Birth*, Robin Reid,[13] associated for many years with the International Livestock Research Institute in Kenya, chronicles research that indicates varying responses of savannas to grazing by wild and domestic herbivores, depending on a suite of ecological factors. For example, multitudes of large wild herbivores have grazed Africa's savannas for millions of years. Their herbivory preadapted their rangelands to livestock. Rangelands adapted to herbivory by fewer and smaller herbivores (impala, gazelles, oribi, warthog, browsers) may be more sensitive to grazing by cattle.

Responses of savannas to high and low livestock grazing intensity can be judged by productivity as measured by the number of plant or animal species. Surprisingly, some savanna grasslands show no response, notably, the short grasslands of very dry savannas like the Serengeti plains. A *simplifying response* may be the reaction to even moderate grazing of savannas poorly adapted to heavy grazing; such land can be degraded to the point of no return. In wetter savanna, as in Serengeti medium and long grassland, pastures may respond to moderate grazing with greater productivity but become less productive with increased utilization; the plotted curve of increasing then decreasing productivity is known as a *humped response*. Finally, there is the *enriching response* to intense livestock grazing of savannas long adapted to heavy grazing that actually degrade when ungrazed.[13] In general, the more abundant water and nutrients are, as in savannas with higher rainfall, the more resilient plants are to grazing. Consistent grazing by wildlife, like fire, keeps grass growing and in its most digestible state. Although such pastures are also resilient to livestock, even the best can be degraded when settlements are permanent and livestock are herded over the same ground month after month without any rest of the range; ground cover and erosion are the result. Pastoralism based on transhumance is the closest approximation to the movements of wildlife and the least damaging to the habitat.

Why, then, are arid and semiarid grasslands not degraded by heavy utilization? In fact, such regions are more subject to droughts that kill off large numbers of livestock and even wild herbivores. Maybe that's the answer, as Reid suggests.[13] For at least several subsequent years the range is rested while livestock numbers recover. In such ecosystems there is little feedback or coadaptation between livestock and vegetation. This implies that livestock herders can have as many animals as they want in dry savannas without worrying much about degrading the habitat or lowering productivity. But again, this holds true only where human population remains relatively low and livestock is free to range over a wide and ecologically diverse area. Degradation increases and may become irreversible when transhumance is replaced by settlement, farming, and increasing population density.

Reid also argues that wild herbivores are attracted to human settlements, coming near but not too near at night. The grass is kept short, making for better grazing and increased safety from predators. I have seen that in the Masai Mara, wildebeest, zebras, and gazelles seem to be more common in the vicinity of Masai settlements. At a lodge in Tsavo East National Park, I was surprised to find a herd of impala lying undisturbed within yards of the path to the dining area and to learn this was a nightly event.

Furthermore, the bomas where pastoralists' livestock are kept at night so enrich the soil that these "hotspots" produce the most nutritious available grazing for possibly centuries after abandonment. (More on this subject in chapter 9.)

SAVANNA GRASSLANDS

As we have seen, natural grasslands exist under soil and drainage conditions that discourage woody growth. Five kinds of natural African grasslands have been described.[10,14]

Small valley grasslands, found mainly in regions of relatively abundant rainfall, especially in the *miombo* woodland, where they are called *dambos* or *vleis.* Depending on size, annual rainfall, soil structure, etc., some have a seasonal or perennial stream in the middle, others retain swampy places, and still others dry out completely. But all are waterlogged by seepage for at least part of the year, so that even the sloping sides, though not actually flooded, stay too wet for trees. The soils are generally leached, slightly acid and infertile in the wetter parts of the moist savanna. The grasses tend to be fine, though tall *Hyparrhenia* may also occur. In *dambos* that retain sufficient moisture through the dry season, a variety of sedges and flowers grow.

Floodplain grassland. On flat or gently undulating plains, the combination of seasonal flooding and protracted drought creates conditions too harsh for all but a minority of specially adapted trees—mainly acacias. Many acacias are adapted to protracted drought, but few can withstand protracted flooding. Some palms and acacias may grow on slightly elevated ground, while trees and bushes may be numerous in places that are only flooded occasionally. A large variety of grasses grow on the floodplain. Most grow tall and rank, but where the water is brackish, specially adapted species predominate. The latter are usually short and fine, and the soil is lighter textured and better drained. This holds particularly for the alluvial plains around the alkaline lakes found in troughs and closed valleys of the Rift Valley, such as Lake Natron.

Farther south, very extensive floodplain grasslands are found in Zambia along the Upper Zambezi and its tributaries in Barotseland; around the Busango and Lukanga swamps and the Kafue Flats in the Kafue basin; the Luangwa Valley; the grasslands bordering the Okavango, Linyanti, and Chobe swamps; and the Etosha Pan in Namibia. Every one of these plains is renowned for great concentrations of wildlife. These and a few other alluvial plains are the only important areas of open grassland in the *miombo* and bushveld zones. Smaller alluvial grasslands border the lower courses of rivers draining into the Indian Ocean through the coastal lowlands of Mozambique and Tanzania.

Cotton-soil grassland. Black-cotton soil is very dark, heavy clay that dries brick-hard, cracks badly in the dry season, and expands greatly when wet. Once saturated, driving and even walking on it is maddeningly difficult. Heavy clay soil (vertisol) of this type is very common in arid savanna, where it covers gentle valleys and even wide plains. It is similar to alluvial soil but is actually illuvial, that is, derived from accumulated fine detritus washed down from surrounding higher ground. Whereas *dambo* soils contain too much water, cotton soils provide too little for most plants, absorbing so much moisture during expansion that hardly any is available to roots. This becomes evident early in the rainy season, for cotton-soil vegetation lies dormant long after plants in better-drained soils have sprouted. At the opposite extreme, heavy rain quickly puddles the surface, sealing it and creating pans that only slowly evaporate or drain from the flat ground. The associated grasses are usually tall and coarse, compared to the species growing on better-drained sites. Gall acacias (a.k.a. whistling thorn) are adapted to these conditions and often form fairly close scrub woodland on what would otherwise be pure grassland. This soil/vegetation association is called *mbuga* in East Africa. Typical examples are seen on the Masai Steppe, where *mbugas* are practically the only open grassland in country covered with thornbush.

Naturally short grassland occurs in arid environments where hardpan forms at a shallow depth due to the accumulation of unleached minerals as an impervious layer of salt-impregnated clay, through which water and plant roots are unable to penetrate. The same conditions are created by a layer of subsurface ledge rock. The B horizon of hardpan soils is generally completely base-saturated and frequently alkaline, conditions that are inimical to most plant growth. A distinguishing characteristic of this type of grassland is that it does not become waterlogged; the soil is permeable enough so that the surface generally remains dry. It is no coincidence, therefore, that the most extensive grasslands of this type occur on sand or equally porous material: namely, the Kalahari sands of southern Africa and plains overlain by deposits of volcanic ash in East Africa.

The foremost examples are the eastern Serengeti plains of Tanzania, characterized as microperennial grassland,[15] the short-grass plains bordering the Makarikari Pans of Botswana, and Namibia's Etosha Pan. Though the soils of the last two are doubtless alluvial in origin, flooding has been rare for a long time. The floodplain grasslands surrounding alkaline lakes resemble typical alkaline short grassland despite their periodic flooding.

Subdesert grassland occurs where there is insufficient water to support trees—the opposite of the waterlogging that keeps trees from growing on alluvial plains. Bushes and trees are confined to drainage lines, while grasses persist on drier terrain. However, the grasses in such an environment are sparse at best. Also, there are many areas where bushes and dwarf shrubs tolerate greater aridity than do grasses, for reasons that remain obscure.

Upland grasslands or downs, found in temperate and subtropical climates, occur only at high altitudes in tropical Africa, especially on plateaus above 6,000 ft. (1,829 m). Isolated plateau grasslands stretch like giant steps from the Ethiopian Highlands to South Africa's Drakensberg Range. A combination of factors interact to produce upland grassland, including cold, wind, chemically rich soil (too rich for most savanna trees)—and again seasonal waterlogging. In areas with high rainfall and frequent mist, soils on slopes may remain too moist for trees even though not actually waterlogged.

The South African highveld represents the lowest grassland of this type, occurring down to 4,000 ft. (1,219 m) and even lower in some places. But unlike upland grasslands, the highveld is temperate grassland that grows in areas with as little as 15 in. (381 mm) of rainfall. Frost, dryness, and strong winds rather than excess moisture are considered responsible for the absence of trees.[16]

SO FINALLY WE GET TO THE QUESTION, WHERE DOES THE WILDEBEEST FIT IN?

The wildebeest inhabits savannas with rainfall averaging between 19 in. (400 mm) and 32 in. (800 mm), spanning arid to moist, or mesic, savanna. A quintessential plains antelope, it is a pure grazer with a preference for short pastures and is closely associated with acacia savanna. Migratory populations disperse over the arid part of the range during wet seasons and concentrate in higher rainfall areas with permanent water during dry seasons. It is rarely found above 6,000–7,000 ft. (1,800–2,100 m) but may be transient in montane grassland and hilly terrain while moving between seasonal pastures.[17,18]

The wildebeest's geographic range is limited to eastern and southern Africa, from just south of the equator in central Kenya, through all the countries of the subcontinent except the Democratic Republic of Congo. The Orange River marks the southern limit of its range and the transition of tree savanna to the temperate highveld. In precolonial times, the common wildebeest dominated the teeming assemblages of plains game in most of the acacia savanna ecosystems of eastern and southern Africa. Loss of range outside of protected areas has resulted in the replacement of migratory populations with smaller, more sedentary populations within protected areas. Wildebeest have also been introduced to regions outside their former distribution range, including the Eastern Highlands of Zimbabwe and private farmland in Namibia.

The history of the migratory populations in colonial and postcolonial times is chronicled in Estes and East[19] and summarized in chapter 4.

GRASSES AND NUTRITION

The characteristics of grasses singled out below have a direct or indirect bearing on wildebeest grazing preferences and will help to explain the distribution and seasonal movements discussed in later chapters. (N.B. I wrote this overview in 1967 when I started—then shelved—a book on the wildebeest as my dissertation, as noted in the introduction. Fortunately the information is still accurate.)

Sweet and sour grassland. The terms *sweet grassland* and *sour grassland* (translated from Afrikaans) refer to nutritional quality and palatability, and perhaps to the fact that the soils of "sour" grasslands are usually slightly acid. Generally speaking, the tall grasses of moist savannas are "sour," while the grasses of arid savannas are "sweet." Early in the growing season, both are highly nutritious, perhaps equally rich in proteins and minerals. But during the dry season the nutritive values of sour grasses fall so low that domestic stock loses weight unless given supplementary feed. For cattle, a crude protein content of 7 percent is considered minimal for body maintenance. The protein content of many sour grasses may fall below 2 percent, compared to 14 percent or higher early in the growing season. In South African highveld, sour grassland provides adequate nutrition for cattle for only three or four months of the year.[16]

What makes grass "sweet" is (a) quick-drying, so that translocation of nutrients to the roots is prevented and proteins and minerals are trapped in the leaves and stems; and (b) a relatively low content of lignin and cellulose, strengthening elements that are needed to support taller grasses, which make grasses harder to digest. Naturally short, fine grasses that cure well on the stem are characteristic of semiarid and arid savannas. With a rainfall of above about 30 in. (800 mm) and/or on seasonally waterlogged soils, the process of translocation goes unchecked until most plant nutrients are stored in the roots, while increased production of cellulose and lignin makes the stems and leaves fibrous and woody. At this stage, grass is not only unpalatable but also largely indigestible for most grazers. In addition, the leached soils typical in regions of high rainfall, particularly on senile granite soils of the plateau (e.g., in Zambia)[20] are generally deficient in essential minerals such as potassium, phosphorus, and calcium, which are required by both grasses and grazers.[12,17,21]

Wherever there is a definite growing season, grasses are most palatable and nutritious in their early stages, when they contain the highest concentrations of proteins and minerals and the lowest amounts of silica, cellulose, and lignin.[22] While most grasses are adapted to being grazed, they are susceptible to damage at certain stages of growth, especially very early, at flowering and seed maturity, and near foliage maturity. Early growth in perennial grasses is produced from food stored the previous year in the roots; the plant makes 15–25 percent of its growth before enough food for its needs is manufactured in the leaves.[23] If the plant is grazed (or burned) before this stage, regrowth occurs at the expense of the already partially depleted root store, and the second crop will be less vigorous. If close grazing continues, the grass may starve and the pasture will be seriously damaged. At flowering and at seed maturity, the proteins, carbohydrates, and fats are largely utilized for regeneration, leaving little in reserve in case of defoliation. The flowering and fruiting heads are the most nutritious part of the plant. Finally, late in the rainy season at foliage maturity, when most of the food that is manufactured is being translocated to the roots, grazing or burning can harm early growth in the following rainy season, both by preventing food storage and by stimulating new growth that depletes existing reserves.

Some grasses have a remarkable ability to withstand almost continuous close grazing, while others can stand very little. The familiar creeping lawn grass *Cynodon dactylon,* or Bermuda grass, and the even better *Cynodon plectostachys* are outstanding forage plants, maintaining food reserves at a high level despite close grazing. So much food is stored in their rhizome system that only a small part is used even in starting new growth.[2] But many, perhaps most, grasses do better with some grazing than with none, at least during vigorous growth. Wellington[16] points out, "Unless the grass is heavily grazed when it can best stand it, the tendency is for more palatable species to be eliminated by selective grazing and the ranker types to take possession of the veld" (294).

In South Africa it is now recognized that sour veld should be heavily grazed during the summer rains and that sweet veld should be reserved for winter grazing—which is precisely what wild herbivores have always done. "For grasslands to be ungrazed is as unnatural, and may be as detrimental to some species, as too much grazing," writes Heady (37).[23] One reason is that the seedlings of short and medium-length grasses, which frequently grow as an understory to tall grasses, may be eliminated by accumulated litter. Heavy grazing serves to both crop and trample the high grass and litter.[2] Second, some good forage grasses, of which Bermuda grass is one, are shade-sensitive and tend to die out under tall grasses. Although it is the least satisfactory alternative, burning is now considered necessary to the management of South African sour veld in the absence of grazing or mowing; not long ago, burning was universally condemned by experts as leading to soil erosion.[16]

Vegetational changes brought on by cessation of burning may be more detrimental to plains game than burning.[23] New growth on burned areas is universally preferred to new growth on unburned areas, although equally green and nutritious new shoots may lie beneath the old growth. The admixture of unpalatable dead stems with the new grass accounts for the preference for the unadulterated new growth on burns.[16, 23] The best time to burn is early in the growing season, after the first rains have fallen, so that the soil is moist and the stimulus to the roots to produce new growth is sustained by the soil moisture.[16] Grass is more nutritious than all other vegetation in its early growth stages and is then preferred by nearly all grazers. Young, tender, rapidly growing grass may contain well over 30 percent crude protein (on a dry weight basis), whereas legumes in the same sward contain 20 percent or less.[2]

The widespread notion that grasses are higher in carbohydrates and lower in protein than the legumes (e.g., clover) is based on the practice of harvesting hay only at maturity and of growing grasses on the poorest, and legumes on the best, soils. Nevertheless, browse forage and legumes in particular are subject to less seasonal variation and are far richer in protein, minerals, and water in the dry season. The legumes are then generally the richest source of protein and are less fibrous than the grasses and browse.[24] Moreover, legumes and browse contain only a third as much of the silica in grasses that wears down grazers' molars. The richest sources of calcium and phosphorus are nonleguminous browse; leguminous browse tested the lowest in phosphorus content. But when all is said and done, new grass may nevertheless be the most nutritious of all forage.

From a nutritional standpoint, the grazing is best from late in the growing season to early in the dry season, for this is when the greatest weight gains are registered by livestock. Although the proportion of crude protein and minerals is lower at this season, there is more grass, so that the yield of crude protein increases even though the percentage in the forage drops. The timing of the wildebeest rut (chapter 11) is linked to the season of maximum grass productivity. The protein yield of mature grass continues to be adequate well into the dry season. In addition, the time of maximum dry-weight production varies widely between different grasses, which results in a long period during which animals can find nutritious feed by selective grazing.[23]

Many tropical grass species that produce very high yields are low in protein after the first flush of growth. Any plant that suffers from an inadequate nitrogen supply will be low in protein, and it is very difficult to supply sufficient nitrogen for some of the rapidly growing tropical and subtropical grasses.[2] Anderson writes,[25] "Without active, nitrogen-producing legumes, the optimum productivity of pastures will never be realized, short of applying heavy dressings of nitrogenous fertilizers" (217). Pastures utilized by large numbers of wild herbivores receive the necessary heavy dressings in the form of manure and urine. Thus, Heady[23] writes, "The contrast is striking in the condition of the land resources between areas used by livestock and by the natural wild fauna consisting of many species, each in its particular ecological niche. Most game areas, but not all, are in excellent condition. Few livestock areas are satisfactory" (37).

Considering the research of Reid,[13] supported by the findings of other pastoralist scientists,[26] readers should bear in mind that livestock do not always degrade savanna ecosystems. Under favorable conditions, livestock can actually enrich rangelands and improve conditions for associated wildlife.

REFERENCES

1. Conway 2009.
2. Harlan 1965.
3. Scholes and Archer 1997.
4. Thompson 1965.
5. Nicholson 2000.
6. White, 1983.
7. Trapnell and Langdale-Brown 1962.
8. Brown 1965.
9. Bond, Woodward, and Midgley 2005.
10. Bond 2008.
11. Sankaran et al. 2005.
12. Sinclair et al. 2008.
13. Reid 2012.
14. Michelmore 1939.
15. Anderson and Talbot 1965.
16. Wellington 1955.
17. Estes and Small 1981.
18. Musiega and Kazadi 2004.
19. Estes and East 2009.
20. Darling 1960b.
21. Kreulen 1975.
22. Dougall, Drysdale, and Glover 1964.
23. Heady 1960.
24. Dougall 1963.
25. Anderson 1965.
26. Homewood and Rodgers 1987.

CHAPTER 3

Introducing the Wildebeest's Tribe

Similarities and Differences among the Four Genera and Seven Species

The Alcelaphini include seven living species in four genera:

Alcelaphus buselaphus, hartebeest

A. lichtensteinii, Lichtenstein's hartebeest

Damaliscus lunatus, topi/tsessebe/tiang

D. dorcas, blesbok/bontebok

Beatragus hunteri, hirola/Hunter's antelope

Connochaetes taurinus, blue/white-bearded wildebeest/gnu

C. gnou, black wildebeest/white-tailed gnu

A wholly African tribe or clade (= branch) of the bovid family tree in origin and distribution, alcelaphines reached their peak two million years ago, when there were over eight genera with at least fifteen species. Though now pared to four genera and seven species, this tribe remains one of the most numerous and widely distributed of African bovids.

TRIBAL TRAITS

The tribe comprises antelopes of medium to large size (blesbok, 60 kg [132 lb.]; blue wildebeest, 237 kg [522 lb.]) with high shoulders sloping to hindquarters, short necks, long thin legs, and horns of similar shape

FIGURE 3.1. a) Adult male hirola, relict species replaced by hartebeest and topi.
b) Territorial Coke's hartebeest (Serengeti NP). c) Lichtenstein's hartebeest, inhabitant
of *miombo* woodland. d) East African topi, territorial male standing on termite mound
(Serengeti NP).

FIGURE 3.1. *Continued* e) Territorial male bontebok, Cape Province, RSA. f) Adult male black wildebeest (a.k.a. white-tailed gnu), South African highveld/karoo. g) Territorial male western white-bearded wildebeest, Ngorongoro Crater. Photos 3.1 (a), (c), (e), (f) courtesy of Ken Coe.

in both sexes. Both sexes have functioning preorbital glands below their eyes and scent glands between the hooves of their forefeet (vestigial in the hind feet). They have only two teats. The calves of all species are tan and look a lot alike.

HOW THEY MOVE

Fleet and enduring runners, topi and hartebeest were the only two antelopes Teddy Roosevelt's African hunting party couldn't run down on their Western cowponies.[1] Alcelaphines are also capable of moving long distances at an easy canter, facilitated by their overdeveloped forequarters and spring-loaded pasterns[2]. Trotting is mainly performed as a display of excitement or alarm. Hartebeest and topi, but not wildebeest, also stot. All species kneel when fighting and to rub their heads in mud. Wildebeests are the only antelopes known to roll onto their backs during the kneeling and horning display.

There is a clear division in the tribe, with *Alcelaphus, Damaliscus,* and *Beatragus* on one side and *Connochaetes* on the other. Hartebeest, topi, and blesbok have narrow muzzles; lyrate or complexly recurved, ringed horns; and short, glossy coats of plain tan to chestnut with or without a light-colored rump and contrasting markings. Their tails are hock length, with a "toothbrush" terminal tuft (longer and white in the hirola). Both in appearance and behavior they are clearly more closely related to one another than to the wildebeests (see fig. 3.1).

The hirola or Hunter's antelope seems to combine traits of both hartebeest and topi. Fossil remains show the hirola was once distributed in eastern and southern Africa. Its current range is limited to southeastern Kenya and southern Somalia, where it is critically endangered[3]. The hirola may represent a less specialized ancestral form that the topi and hartebeest have replaced within the last million years, possibly as a result of a wetter stage between Ice Ages. (Grevy's zebra has a similar history.) The hirola overlaps with the topi in the savanna between the arid *nyika* thornbush and the coastal forest-savanna mosaic. It does not overlap with the hartebeest, which appears to be its closest competitor, except for a small introduced hirola population in Tsavo NP.[4]

Wildebeests have a broad muzzle, flexible lips, unringed horns hooked forward or sideward, with knobby bosses. The coat is short and glossy, adorned with mane, beard, and facial tufts. The pelage is dark brown to black with a black or white tail (white-tailed gnu) to blue-gray with black stripes and countershaded, darker below and lighter above (blue wildebeest). The tail is long and flowing, as in horses.

Similarities in size, conformation, and movements of topi and harte-beest can make it hard to tell them apart at a distance or in poor light. But near at hand you can see striking differences. Take the horns, for instance. The topi's horns are simple in shape and size compared to their maximum development in lelwel and red hartebeest *(A. b. lelwel and caama)* (fig. 3.2). The bony pedicel between the eyes and the horn base is also most developed in these subspecies, adding inches to the elongated skull.

Coloration and markings make for even more obvious differences between the two genera, and the contrast is most pronounced where the two overlap. Yet their color patterns vary regionally. In Tanzania and Kenya, Coke's hartebeest *(A. b. cokei)* is plain tan with a white rump, whereas the topi *(D. l. korrigum)* has conspicuous purple blotches on the forequarters and upper limbs that contrast with tan lower legs, a black forehead, and reverse countershading (lighter on top). In southern Africa, the red hartebeest *(A. b. caama)* is the gaudy one, with markings rather like the topi's in East Africa (fig. 3.2). The tsessebe's *(D. l. luna-tus)* coloration looks like a washed-out version of the topi, and the horns are much weaker, in striking contrast to the very prominent horns of the red hartebeest.

The concept of character displacement helps explain how such differences evolve.[5] Consider this scenario: Two populations of a given species become separated (as during the Ice Ages) and over time diverge enough to be considered subspecies. When or if climate change reconnects the subspecies, they may freely interbreed, leading to the disappearance of the subspecies differences. But if accumulated differences in genotype and phenotype (appearance) have created barriers to hybridizing, then natural and sexual selection will accentuate their differences. When two subspecies coexist without interbreeding, they have become two different species.

In the case of hartebeest and topi, the differences are sufficient to consider them different genera, but at what point interspecific differences reach this higher category is in the eye of the taxonomist. Two species cannot fill the same ecological niche; one is bound to supplant the other, according to the rule of competitive exclusion—unless they diverge. So selection will operate to avoid competition by making them

FIGURE 3.2. a) Red hartebeest of the Southwest Arid Zone, with pronounced markings more like a topi than an East African hartebeest. b) The tsessebe of the Southern Savanna is a faded, weak-horned race of the topi. Photos by Ken Coe.

more unalike, in size, color and markings, habitat, and diet. Sexual selection favors the most contrasting characters. Particularly when there is a close relative present, characters that distinguish one's own species and prevent mistaken identity are strongly selected.

Yet adaptive divergence can only go so far when species or genera have a common ancestor and share the same genetic tool box of possible variations. Topi and hartebeest seem to be more aware of one another than of more distantly related antelopes. They are more often found in mixed herds and interact more often than would be expected by chance. The same is true of Grant's and Thomson's gazelles. Shared ancestry is revealed by occasional hybridization of blesbok and hartebeest on South African farms, even though the contrast in appearance could hardly be greater: the blesbok is arguably the gaudiest of all African antelopes (see fig. 3.1).

GEOGRAPHIC RANGE

The hartebeests had perhaps the widest distribution of any antelope, found in nearly all the grasslands and savannas of Africa. There was even a subspecies north of the Sahara in Algeria, the bubal *(A. b. buselaphus),* which became extinct in 1925.[4] Geographic differences were so marked that eight to ten subspecies were recognized, reflecting adaptations to climate and habitat differences.

The topi also had a very extensive but patchy geographic range, reflecting its specialization for particular grassland communities in the arid and savanna biomes. The tsessebe of southern Africa is separated by hundreds of miles of *miombo* woodland from the nearest topi populations of East Africa. Wherever it overlaps with wildebeest, the wildebeest is dominant. But in the Northern Savanna, the topi fills a similar ecological niche and is a keystone grazer of vast floodplains. Up to three quarters of a million tiang *(D. l. tiang),* the largest subspecies, still shares dominance of southern Sudan with the white-eared kob.[6]

The blesbok and black wildebeest, along with the springbok and quagga (the southernmost subspecies of zebra), thronged the South African highveld in uncounted thousands. That was before settlement, fencing, and market hunting transformed this temperate grassland biome in the late 1800s. The quagga was the only highveld species to become extinct; the other species have recovered in large numbers. But there is no longer

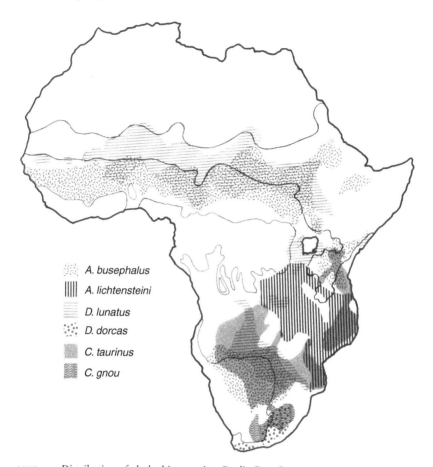

MAP 3.1. Distribution of alcelaphine species. Credit: Dan Otte.

a migration, fences subdivide the populations, and there are no predators bigger than a jackal.

Common wildebeest. Fossil evidence indicates that *C. taurinus* lived in the Saharan region in the late Pleistocene[7] and more recently in West Africa, possibly in a "bovid" western refugium.[8] Whatever caused those populations to go extinct, its historical range has been limited to the savannas of eastern and southern Africa. The northern limit is in central Kenya just south of the equator, and the Orange River, the dividing line between tree savanna and the treeless highveld of South Africa, marks the limit of its southern range.

ECOLOGICAL SEPARATION

Hartebeest. Hartebeest are more tolerant of wooded and bush grassland than other members of the tribe. They tend to be found on the edge rather than in the middle of the open plains. In East Africa, Coke's hartebeest is particularly associated with medium-length bushed grassland growing on black-cotton soil dominated by red-oat grass *(Themeda triandra)* and gall acacias *(Acacia drepanolobium)*. Lichtenstein's hartebeest is adapted to the Miombo Woodland Zone, where it exploits the drainage-line grasslands that dissect the woodland. Hartebeest are among the purest grazers, adapted to feeding selectively in medium-height grassland, and most populations are able to find the grasses they prefer at different seasons without moving long distances. But in some areas, notably Botswana, hartebeests are nearly as mobile as wildebeests, trekking to areas where thunderstorms stimulate a flush of new grass during the long dry season. The species is considered water-dependent, yet red hartebeest can be found in waterless areas of the Kalahari. They must share the ability of desert-adapted antelopes like the oryx and springbok to minimize water loss by physiological and behavioral means, but how they get sufficient moisture without browsing is puzzling. Most of the Kalahari ungulates are known to eat tsama melons and dig up tubers, but these are not always abundant.

Topi. The topi, with a narrow muzzle and mobile lips adapted for selective grazing, eats virtually nothing but grass, harvesting the greenest, most tender blades and avoiding mature leaves and stems. It has an advantage in pastures with old and new growth intermixed but is less efficient than a bulk grazer like the wildebeest on a short sward. Topis in green pastures can go without drinking but drink every day or two when the grass is dry. The densest populations occur where green grass is available the whole year, as on floodplains bordering rivers, lakes, and swamps. Where the distance between wet- and dry-season pastures is many miles, as in southern Sudan, populations are migratory. On plains in the western Serengeti, topi often stay together in megaherds of over 5,000 animals.

Blesbok. In the South African highveld, the blesbok filled a niche for a selective grazer comparable to the topi in the tropical savanna. It was more water-dependent and less efficient on short grass than the codominant springbok and black wildebeest, which ranged the arid karoo in the summer rainy season and the highveld grasslands in the dry season.

The blesbok spent the wet season on the highveld and went east into the taller, wetter grasslands (the so-called sourveld) during the dry season, where it could pick and choose nutritious growth more efficiently than the wildebeest. Even so, it needed new growth stimulated by wildfires to stay in good condition.[9,10]

Wildebeest. Wildebeest are equipped to take big bites of short green grass; a blunt muzzle and wide incisor row, as well as large body size, prevent their feeding efficiently on widely separated plants. Accordingly, they favor short mat-forming grasses like *Sporobulus, Digitaria,* and *Cynodon* spp. that carpet short-grass plains, where thousands of animals grazing close together can eat their fill. Short grasslands of this description grow on alkaline and volcanic soils in arid regions underlain by a hardpan where only shallow-rooted grasses can grow. Grazing, trampling, and manuring by thousands of animals stimulates new growth—as long as the ground holds moisture. The Serengeti and all the other great migratory wildebeest populations thrived on these grasslands during the rainy season and survived on longer grasslands in areas of higher rainfall and permanent water during the dry season. Man-made wildfires and occasional thunderstorms enabled part of the population to enjoy green pastures at the most nutritionally critical time of year. Grasslands bordering alkaline lakes and pans, where colonial and annual grasses carpet the emerging ground, are also preferred dry-season pastures. A pure grazer that has to drink every day or two in the dry season, the wildebeest cannot use pastures beyond convenient commuting distance of water, say, 16 to 25 km (10–15 mi.). Yet like red hartebeest, some Kalahari wildebeest survive in waterless areas. On the other hand, the wildebeest's ability to lope mile after mile enables it to travel up to 76 km (50 mi.) in a day to reach a green flush brought on by local showers or burning.

The different habitat preferences of the alcelaphines are obvious enough. But what are the mechanisms that gave rise to the separation of associated species? Some possibilities that ecologists have investigated are dietary specialization, competition between species, facilitation between species, niche overlap, and migration. Bell[11] and Jarman[12] concluded that differences in body size and digestive system are most important in niche separation.

Martyn Murray[13] compared the niche separation of wildebeest, topi and hartebeest by measuring bite weight, bite rate, intake rate, and selectivity of tame animals kept at the Serengeti Wildlife Research

Center in plots containing grass at different growth stages. Close observation filled in details of the feeding and digestive mechanisms that make for niche separation.

1. On growing swards, hartebeest had a smaller bite weight and lower intake rate and were also less selective of green leaf than either topi or wildebeest. On mature swards, hartebeest were more selective of leaf than the other two species.

2. Wildebeest had a faster bite rate than either topi or hartebeest on swards with low biomass and high protein content of green leaf (green flush).

3. Topi were significantly more selective of green leaf than the other two species and were the only species to maintain a rapid bite rate on swards with high green-leaf biomass.

4. Each species was most proficient either in leaf selection or bite rate when feeding on grass swards in its preferred growth stage.

Murray and Brown[13] concluded that pasture growth stage was a primary determinant of niche separation. In the Serengeti, the three dominant migratory species, wildebeest, zebra, and Thomson's gazelle, are specialists of the earlier growth stages of the grasslands, which tend to be transient. The resident grazers, notably hartebeest, waterbuck, reedbuck, and buffalo, specialize on later growth stages of the longer grasses as they cure to standing hay.

In a separate feeding trial, in which selectivity, consumption, and digestion of equal amounts of chopped grass by the tame antelopes were analyzed, hartebeest proved to be more selective and to digest fiber more efficiently and in smaller amounts than wildebeest or topi.[13] In short, the hartebeest eats less but feeds more selectively and digests fiber and organic matter more efficiently than the other two species. This enables the hartebeest to subsist on low-quality forage by selectively feeding at a low daily intake compared to its two relatives. Murray and Brown suggest that its frugal appetite may be linked with a comparatively slow growth rate.

TERRITORIAL MATING SYSTEM AND SOCIAL ORGANIZATION: THEME AND VARIATIONS

In territorial mating systems, exclusive ownership of a piece of real estate is necessary for a male to have the opportunity to reproduce.

Frequency of mating opportunities depends on where the territory is located and on female herd size (population density). Males therefore compete for territories in the best habitat. The fittest males hold central locations. Competition for choice spots is intense, but once established, ownership confers such a psychological advantage that serious challenge by an outsider is rare. The owner maintains the status quo by scent-posting the property and engaging in daily ritualized encounters with his territorial neighbors. Wildebeest territorial behavior is detailed in chapter 8.

Although actively territorial males seldom constitute more than 10–20 percent of a population, they enforce sexual segregation. Intolerant of the intrusion of other males, they exclude from female herds all those displaying obvious male secondary characters and sexual behavior. As a result, females and young are contained within the mosaic of individual territories, while nonterritorial males end up in bachelor herds in less desirable and actively defended habitat.

As discussed in chapter 1, the lack of conspicuous sexual dimorphism in this whole tribe can be explained as the consequence of mothers mimicking male secondary characters, including horns, as a means of postponing the eviction of sons until they can cope with leaving the maternal herd and home range. (See full presentation in Estes.)[14]

Territoriality is highly variable. Adaptability is the reason it is so widespread among antelopes. The forms it takes are an integral part of a species adaptation for a particular ecological niche. In a sedentary-dispersed population, males defend large, resource-based territories in a resident population of small herds composed mainly of related females and separate herds of bachelor males. In a mobile-aggregated population, males and females intermingle as long as the aggregation keeps moving. During pauses, males immediately space out, establish small temporary territories, and proceed to cut out the noncompeting males. The range of variation between these two states is clearly illustrated in the alcelaphines. (See tribal and species accounts in Estes[10] for details such as territory size and behavioral comparisons.)

Hartebeest

In keeping with its frugal use of resources and preference for medium grassland along woodland edges, most hartebeest populations are able to find their seasonally preferred feeding grounds without moving long distances. A typical arrangement, as seen in the upland grassland of

Nairobi National Park, is for the herds to concentrate on the hilltops during the rains, where they maintain grazing lawns of nutritious green grass, then to move downhill into longer, still-green grass during the dry season, ending up in the tall, coarse, still-green long grass in clay soils at the bottom of the catena in the late dry season (see fig. 9.2). The best territories include a swath of all the grassland types; these are permanently occupied but only by the fittest, prime males. Older males remain territorial but end up on less desirable locations on black-cotton soil among the whistling-thorn scrub, where females are rarely found.[16]

Topi

Topi social organization is perhaps the most variable of any African bovid. The form it takes depends on the nature of the habitat and associated population density. Territoriality adjusts accordingly. In mixed woodland and grassland, the prevailing pattern is small, semiexclusive female herds whose range is divided among a small number of permanently occupied territories of up to 2 sq. km (0.77 sq. mi.). Extensive open grasslands are occupied by aggregations numbering from hundreds to thousands, including actively competing males and noncompeting bachelors. In Uganda's Queen Elizabeth National Park, well outside the wildebeest's East African range, there used to be a population of around 1,000 topi on the Ishasha Plain. These animals were not migratory but nomadic: the plain was a patchwork of different grassland types, and the topi, often in a single, dense aggregation, moved from patch to patch. Meanwhile, on the other side of the Ishasha River, in a different habitat with more trees, was a small resident population of separate female herds and spaced territorial males.

But what puts the topi over the top is lek breeding. Instead of defending a conventional resource-based territory, a bunch of males cluster close together on a dedicated display ground or arena. During the annual mating season, large aggregations surround a display ground, which may be used for many generations and is located on the widest area of grassland, especially where plains intersect. Females visit the lek the day they come into estrus and end up mating with one or a few centrally located stud males who, research has shown, are the biggest and fittest of the contenders.[17,18] The topi is one of only four antelopes known to form leks; the other three are the kob *(Kobus kob)*, puku *(K. vardoni)*, and lechwe *(K. leche)*, all members of the Reduncini tribe.

Blesbok

Quite possibly the blesbok was also a lek breeder, but migratory populations were fragmented before anyone studied the species. Under current conditions, the sedentary-dispersed distribution pattern is perforce the prevailing mode.

Black wildebeest

The black wildebeest is another species whose behavior under natural conditions was never studied. Like its congener, it migrated in huge numbers in the ecosystem to which it is adapted: the temperate, nearly treeless steppe of South Africa's highveld. The current population of 20,000 or so lives on fenced ranches and game reserves in the typical sedentary-dispersed mode, with females and young distributed within a network of territorial males and herds of bachelor males kept away from the females by the landlords. Though barely half the weight of a blue wildebeest, resident *gnou* males defend much bigger territories: spaced about one kilometer (0.62 mi.) apart at low density down to a minimum of approximately 180 m (197 yd.). Just how compressed territories might have been in great aggregations is an interesting question, considering that two or more rutting Serengeti bulls can operate around the same tree (chapter 8). The territorial call of the black wildebeest is a high-pitched hic, without the cough, that carries for over a kilometer.[19]

Territorial Advertising

The single most important territorial rule is being present. Absence invites invasion by territory seekers; the longer the absence, the greater the risk. The more conspicuously presence is advertised, the better. It is customary for a territorial male to stay some distance from a herd on his property, preferably in the most exposed location. Topi have a particular preference for termite mounds. Hartebeest also like elevations but are less committed. A territorial male can see and be seen to advantage atop the mound; he advertises territorial status just by standing there with head raised, as nonterritorial males and females of all these species stand with head at or below shoulder level absent something that makes them assume the (similar) alert posture. Meanwhile, his lordship is surveilling for invading males or incoming females as much as for predators and likely chewing his cud at the same time.

FIGURE 3.3. Topi rocking horse canter.

A more obvious territorial display is the rocking canter, which territorial wildebeest, topi, and hartebeest deploy when going to meet approaching conspecifics of either sex. Males galumphing through an aggregation and even standing still stand out from all the others in the normal attitude with head at shoulder level. When meeting females, the males segue into the low-stretch courtship display, with nose out and ears down. The topi's display is particularly impressive because males high-step, accentuating the contrast between dark upper and tan lower legs (fig. 3.3).

Territorial males of all alcelaphine species maintain patches of bare ground where they regularly deposit dung, paw, kneel and horn the soil, and lie down. These stamping grounds are the olfactory center of the territory. Topi and hartebeest often establish their stamping grounds at the base of a termite mound. The dunging ceremony also doubles as a visual display, often emphasized by pawing. Mutual defecation is also a regular feature of the challenge ritual.

Comparing kneeling and horning behavior, topi, hartebeest, and common wildebeest are addicted to mud packing from this posture. Both sexes participate, though it is primarily an expression of an aggressive mood and therefore mostly a male activity, with territorial males the leading actors (also true of wildebeest tree horning; see chapter 9). Wet soil stimulates mud packing. Topi and blesbok cake their horns with mud. Hartebeest coat their muzzles and horns, then wipe the mud on their shoulders and flanks. Pleading guilty to anthropomorphism, I picture topi trying to make their horns more impressive, while hartebeest seek to darken their tan coats, like sunbathers acquiring a tan. Wildebeests horn and also roll in mud; they even roll on their backs, the only alcelaphines to do so.

Both sexes of all species have active, pecan- to walnut-sized preorbital glands that territorial males use to delineate their property, accessory to their dung deposits. In the topi/hartebeest division, the glandular secretion exudes from a central orifice that can be closed and opened like a purse.

Both sexes of hirola, hartebeest, blesbok, and topi (especially) insert grass stems or twigs into the opening and coat them with a gel-like deposit, which they often wipe onto the horns by stereotyped weaving movements. Preorbital secretions are not only species- but also individually specific; that is, they identify the marker, his or her reproductive and social status, and even denote how long ago the deposit was made. If it is a territorial male's mark, visitors who encounter his scent marks will recognize and most likely avoid messing with him. Gosling[20] terms this scent-matching. The owner ensures that his odor is detected by wiping preorbital secretion on his own body with vigorous side-to-side head movements.

Interestingly, the secretion varies even between subspecies. The lelwel hartebeest's secretion is colorless, whereas the secretion of Coke's hartebeest is a sticky black paste containing melanin, which dries as a tarlike deposit. Self-anointing with preorbital secretion is a hard-wired *fixed action pattern*. Lichtenstein's hartebeest of all ages mark the same spots on their shoulders and flanks so consistently that the black patches have been mistaken for species-typical markings (see fig. 3.1).

The wildebeests' preorbital glands are bigger, hair-covered, and ductless. They rub the oily secretion exuded by the glands on trees (see chapter 8), the ground, and one another.

Reproduction

Like all African antelopes, alcelaphines produce a single young. As a rule, mating occurs during or following a time of plenty, when the population is in peak condition, and fertility rates commonly exceed 90 percent. The eight-month gestation is adapted to span the dry season and have calving coincide with renewed rainfall and green pastures. In equatorial regions where rain falls in most months, species that are strictly seasonal elsewhere may have a more extended or even perennial

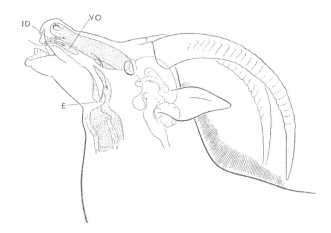

ID

VO

E

FIGURE 3.4.
Schematic of sable
performing
flehmen. Reprinted
from Estes 1991a.
Drawing by
D. Otte.

breeding season, as for example the topi population of eastern Kenya north of the Tana River.[21] As most alcelaphine populations live in regions with dry seasons lasting up to six months, seasonal breeding is the rule. Common wildebeest populations are strictly seasonal, rutting at the end of the rains and calving during the following rainy season. A large majority of the cows are bred and calve during peaks of three weeks. Topi and blesbok have mating and calving peaks of one to two months but mate earlier in the wet season and calve just at the end of the dry season.

Nearly all mammals have a separate, special olfactory organ that functions to detect the reproductive status of females from hormone breakdown products excreted in their urine. Male ungulates routinely test female's urine, and the male then stands with head raised and upper lip curled in a conspicuous grimace called flehmen (fig. 3.4). This behavior serves to suck the urine sample or scent molecules through twin ducts behind the upper incisors and into the tubular vomeronasal organs that lie on either side of the nasal septum. Females of most species will cooperate by urinating on demand. Depending on the result of the urinalysis, the male will either cease and desist or begin actively courting the female.

Unlike all other known members of the bovid family, *Damaliscus* and *Alcelaphus* species do not perform the urine test, nor do females urinate on demand. How males detect the onset of estrus simply by sniffing female genitalia, which they do, is puzzling but has to involve the other general-purpose olfactory organ. Adding to the mystery is the presence of a normal-looking vomeronasal organ (VO) in *Damaliscus* and *Alcelaphus*. The absence of incisive ducts to draw urinary molecules into the VO is the likely explanation. These changes in skull architecture underline the common ancestry of the two genera and their contrast with wildebeest.[22] What could have happened to cause this change is unknown. But it also underlines the common ancestry of these two genera and their contrast with wildebeest.

Unlike *Damaliscus* and *Alcelaphus* species, the hirola urine tests females, gives the flehmen grimace, and, presumably, has a functioning vomeronasal organ. This can be considered additional evidence of its relict status, predating the origin of the other two genera.[23,24]

Hartebeests breed year-round, with one or two peaks during the rains and late in the dry season. Females come into estrus within a few weeks of calving. Consequently, a cow may be followed by a months-old calf, a yearling, and a two-year-old, in that order. This perennial breeding contrasts with the topi's and blesbok's annual breeding. Their young separate from their mothers as yearlings, when the next calf is born. The same is true of wildebeest, although some yearling females may tag along for a second year (see chapter 11).

The difference between the two systems is explained by the adaptations of *Damaliscus* and *Connochaetes* to nomadic and migratory movements in large aggregations.[15,10] The hartebeest retains the ancestral system of all other antelopes (and also all deer family members): newborns go through a hiding stage lasting days to weeks until they develop the agility to flee from predators. In *Damaliscus* and *Connochaetes* the mobility of large aggregations selected for shorter and shorter hiding stages, leading eventually to the complete replacement of the hiding strategy with a system in which calves accompany their mothers from birth. Different stages in the transition are seen in the tribe. The topi floods the predator market by producing most calves within one to two months but retains most of the elements of the hiding strategy, including isolation during and after parturition, a brief lying-out stage during which voiding of wastes requires licking of the calf's anogenital region, and a relatively undeveloped following response. Blesbok calves do not go through a hiding stage but are less precocious and do not follow their mothers as closely as wildebeest calves. The follower strategy is fully developed only in *Connochaetes*. Why and how are summarized in chapter 11.

REFERENCES

1. Roosevelt and Heller 1914.
2. Alexander 1977.
3. Walker 2012.
4. Kingdon 1997.
5. Estes 2000.
6. Hillman and Fryxell 1988.
7. Gentry 1978.
8. Arctander, Johansen, and Coutellec-Vreto 1999.
9. DuPlessis 1972.
10. Estes 1991a.
11. Bell 1970.
12. Jarman 1974.
13. Murray and Brown 1993.
14. Estes 1991b.
15. Estes 2000.
16. Gosling 1974.
17. Gosling 1991.
18. Bro-Jorgensen 2002.
19. Richter 1972.
20. Gosling 1982.
21. Duncan 1975.
22. Hart, Hart, and Maina 1988.
23. Estes 1999a.
24. Butynski 2013.

CHAPTER 4

Wildebeest Subspecies and Status of Migratory Populations

I'm a g-nu, spelt G-N-U,
Call me 'bison' or 'okapi' and I'll sue.
G-nor am I in the least,
Like that dreadful hartebeest,
Oh, g-no, g-no, g-no . . .
G-know, g-know, g-know, I'm a g-nu . . .

—Flanders and Swan

Subspecies

C. t. taurinus, blue wildebeest or brindled gnu. Namibia and South Africa to Mozambique north of the Orange River, from Mozambique to Zambia south of the Zambezi River, and from southwestern Zambia to eastern and southern Angola. Slate blue coat with conspicuous dark stripes, black beard, and upstanding black mane. Shoulder height (Sh.) males 147 (140–56) cm (58 [55–61] in.), females 135 (129–40) cm (53 [51–55] in.); weight (wt.) males 237, 252 kg (522–56 lb.), females 190–215 kg (418–74 lb.).[1,2]

 C. t. cooksoni, Cookson's wildebeest. Restricted to the Luangwa Valley, Zambia. Vagrants occasionally range onto the adjacent plateau and into western Malawi. Browner than other races. Wt. two males 235, 241 kg (518, 531 lb.); two females 219, 224 kg (483, 494 lb.).[3]

 C. t. johnstoni, Nyassa or Johnston's wildebeest. North of Zambezi River in Mozambique to east-central Tanzania, formerly as far north as the Wami River; southeastern Malawi (extinct). Sometimes referred to as the white-banded wildebeest for the pale chevron between its eyes (often absent in the Tanzania population). Sh. 130 cm (51 in.), wt. 227 kg (500 lb.).[4]

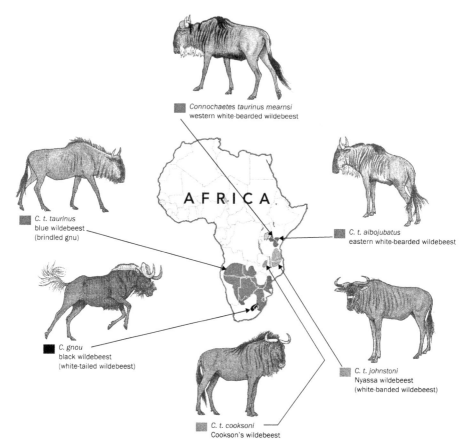

Connochaetes taurinus mearnsi
western white-bearded wildebeest

C. t. taurinus
blue wildebeest
(brindled gnu)

A F R I C A

C. t. albojubatus
eastern white-bearded wildebeest

C. gnou
black wildebeest
(white-tailed wildebeest)

C. t. johnstoni
Nyassa wildebeest
(white-banded wildebeest)

C. t. cooksoni
Cookson's wildebeest

FIGURE 4.1. Wildebeest subspecies and distribution map. By Laura Maestro and Joe LeMonnier, *Natural History*, September 2006. © *Natural History Magazine*. Reprinted courtesy of C. Harris.

C. t. albojubatus, eastern white-bearded wildebeest. Northern Tanzania to central Kenya just south of the equator, west to the Gregory Rift Valley. Southernmost point in recent past (1950s) at least to the Handeni-Kondoa road (5°30′ S).[5] The lightest-colored race, beard white to tan, mane lax and black. Wt. males 243 (222–71) kg (536 [489–597] lb.), females 192 (179–208) kg (423 [394–459] lb.).[6]

C. t. mearnsi, western white-bearded wildebeest. Serengeti-Mara ecosystem of northern Tanzania and southern Kenya west of the Gregory Rift Valley, formerly to Lake Victoria. Smallest race, dark gray or brown, mane lax and black, beard white to tan, horns shorter than in other races but with more developed boss in males. Sh. males 110–34 cm (43–53 in.), females 107–23 cm (42–48 in.). Wt. males 201 (170–242) kg (443 [375–534] lb.), females 163 (141–86) kg (360 [311–410] lbs.).[7]

DNA studies of the different subspecies support the morphological evidence that *mearnsi* stands apart from the rest.[9] Interestingly, *mearnsi* is the only subspecies in which I have observed a faux penile tuft in females. It is developed to a varying degree, as is also the case in males, and is undeveloped in some females. In a 1973 sample of 81 Ngorongoro females, the tuft was developed in 69 and undetected in 12 cows. The tuft is developed in calves by the age of four months old. I interpret the presence of the tuft in female *mearnsi* as a more advanced stage of female mimicry that facilitates peaceful integration of the sexes in this highly migratory subspecies. Selection for development of the tuft would likely center on yearling males: males would be less subject to eviction by beady-eyed bulls if yearling females wore a similar-looking penile tuft.

A much more obvious difference is in the advertising calls of territorial bulls. *C. t. mearnsi* is by far the most vociferous. The eastern white-bearded wildebeest and the blue wildebeest call less frequently, with usually no more than 5–7 grunts in a series. Their voices have a more metallic sound, which the famous South African game warden Stevenson-Hamilton characterized as a "resonant kwank." As shown in the audiospectrograph included in my dissertation publication, the call begins with greatest volume and tails off, whereas *mearnsi*'s call is sustained. Furthermore, the actual vocal mechanisms are different (and unstudied) as all the subspecies but *mearnsi* call with open mouths; *mearnsi* bulls keep their mouths closed.[8] Interestingly, the calls of females and young sound alike in the different subspecies.

SUMMARY OF STATUS BY SUBSPECIES

Blue wildebeest *(C. t. taurinus)*

South Africa

By the 1960s, the blue wildebeest had been shot out over most of its former range in the Kalahari thornveld, bushveld, and lowveld in the northern parts of the country. The major surviving populations were in Kruger NP (13,000), Hluhluwe-Umfolozi Game Reserves (8,000), areas of the Transvaal outside Kruger (7,500, of which two-thirds were in the privately owned game reserves on Kruger's western boundary), Mkuzi Game Reserve (600), and Kalahari Gemsbok NP (resident population of 500, augmented regularly by up to several thousand migratory animals that moved across the border from Botswana). The total population was on the order of 30,000, excluding the migratory animals from Botswana.

Current numbers include 10,000–17,000 in Kruger NP, where the population fluctuates in response to factors such as rainfall and lion predation, and more than 10,000 in total in various smaller protected areas, such as Hluhluwe-Umfolozi Park and other parks and reserves in KwaZulu-Natal, Madikwe Game Reserve, and Pilansberg NP in North West Province. In these smaller areas, removal of animals by harvesting to supply meat to surrounding rural communities and/or by live capture for sale to game ranches is often the major population determinant. The population of Kalahari Gemsbok NP (now part of Kgalagadi Trans-frontier Park) is similar to that in the 1960s. On private land, wildebeest numbers have increased markedly since the early 1980s with the growth of the private game-ranching industry and may now exceed 15,000. There are probably at least 40,000–45,000 blue wildebeest in South Africa at present (excluding migratory animals from Botswana), but this is only a small fraction of the likely numbers that existed in the country 150 years ago.

Swaziland

The wildebeest probably occurred throughout the lowveld region of Swaziland in the past, but the country is now densely populated, and wildlife is restricted to relatively small protected areas. At present, there are probably several hundred blue wildebeest in total in Hlane Game Reserve and a few other protected areas.

Zimbabwe

The wildebeest population of Hwange National Park in the bushveld of western Zimbabwe became established in the early 1930s, when migratory animals from the Botswana border area took up permanent residence around the park's newly established artificial waterholes. This population numbered 1,000–2,000 head in the 1960s and has since remained stable at this level, at least until the mid- to late 1990s.

The other region in which the wildebeest occurs naturally is the lowveld, in southern Zimbabwe. In the 1960s, the lowveld population was at least several thousand and possibly more than 10,000. These animals occurred mainly on privately owned cattle ranches, where bush encroachment was leading to declines in the populations of grazing antelopes. In addition, wildebeest numbers had been reduced greatly through shooting by European farmers, and the then-Rhodesian government was rigorously pursuing a game destruction program for tsetse control even in protected areas. At the time, the prospects for the lowveld's wildlife looked bleak. This was reversed over the next thirty years by the growth of Zimbabwe's wildlife industry on privately owned land. By the mid-1990s, the total number of wildebeest in the country exceeded 10,000, including at least 9,000 head on private land mainly in the lowveld. Since 2000, substantial parts of the country's private wildlife sector have been destroyed by the government's land resettlement program, but wildebeest numbers may not have been reduced greatly because some of the large private conservancies in the lowveld have remained partially or completely intact.

Botswana

The Kalahari Desert, which occupies most of Botswana, the northern Cape of South Africa, and eastern Namibia, formerly supported one of Africa's great plains-game ecosystems. Resident humans were few, mainly Bushmen, and there was not enough surface water to support pastoralism. During the wet season, a large migratory wildebeest population dispersed throughout the Kalahari savannas. When these areas became waterless for several months during the dry season, and especially in severe drought years, the wildebeest concentrated near permanent water in areas such as the Makgadikgadi Pans, the Lake Ngami depression, and the Chobe River in Botswana; the floodplains bordering the Okavango Swamp in Botswana and Namibia; Etosha Pan, Ovamboland, and the Caprivi Strip in Namibia; along the Cunene and Cubango Rivers in Angola; and formerly along the Limpopo River in Botswana and South Africa and the Orange River in South Africa. This vast region may have been occupied by a single wildebeest population, which dispersed widely during the wet season but broke up into separate concentrations during the dry season. Its size will never be known, but it must have comprised at least several hundred thousand individuals.

Much of this system was still intact in the mid-twentieth century, but it has subsequently been destroyed by the erection of game-proof Disease Control Fences in Botswana and Namibia to protect cattle from wildlife-borne diseases. These fences have effectively cut off access between the former wet- and dry-season ranges of the migratory wildebeest. There were still an estimated 260,000 wildebeest in the Botswana section of the

Kalahari in the late 1970s, before access to ancestral dry-season water supplies was finally cut off completely by the fences. Since then the cattle industry has expanded increasingly into the Kalahari through the extensive sinking of boreholes, adding to the pressures on wildlife.

The era of unfettered movement of wildebeest and other game animals across the Kalahari landscape has come to an end. Wildlife is now increasingly restricted to protected areas. The total wildebeest population of Botswana had declined to an estimated 39,000 head in 1986–87 and has subsequently stabilized at about this level; total numbers were estimated to be 35,000–45,000 in 1986–94 and 46,000 in 2002–3.

Like most other wildlife species in Botswana, the wildebeest has fragmented into separate regional populations. The Okavango-Chobe-Makgadikgadi population in the north is currently stable at about 14,000. The erection of additional fences may isolate the Makgadikgadi component, which currently numbers about 4,000, from the population centered on the Okavango. The population of the Tuli Block farms in the east has grown from a declining remnant of a few hundred individuals in the 1960s to about 19,000 at present; this region now has the country's largest number of wildebeest. In central and southwestern Botswana, a more or less stable, remnant population of about 10,000–12,000 animals survives in the central and southern Kalahari; this population is vulnerable to poaching and competition with livestock in areas outside Central Kgalagadi Game Reserve and Kgalagadi Transfrontier Park. There is a much smaller, declining population in the Namibia border region in the northwest.

Namibia

The wildebeest formerly ranged over all but the most arid parts of Namibia. Until the 1950s, it was probably the country's most abundant large herbivore. Former numbers are unknown but may have reached the hundreds of thousands. The eastern and northern regions fell within the vast Kalahari ecosystem (see above). Development of the intensive cattle ranching industry on European-owned farms, ruthless eradication of the species on private land because it carries malignant catarrh, habitat deterioration through overstocking of livestock, and construction of Disease Control Fences have since led to the annihilation of Namibia's wildebeest population. By the late 1960s, numbers had apparently been reduced to fewer than 10,000; by the 1980s, fewer than 4,000 remained.

The main protected-area population is in Etosha National Park. Until the 1960s, the Etosha park supported up to 30,000 wildebeest seasonally, including perhaps 10,000 that moved within the confines of the park and larger numbers that migrated into the park from Ovamboland during the wet season. Following construction of a fence along the eastern border blocking the migration route, the Etosha population had declined to 2,000–3,000 head by the late 1970s and has subsequently remained stable at this greatly reduced level.

The wildebeest has made a small-scale comeback on private land since the 1970s, with the growth of Namibia's private-sector wildlife industry. There are now more than 5,000 head on private farms and conservancies and probably more than 10,000 in total.

Angola

There were formerly substantial populations of wildebeest in the bushveld of southern Angola. The species also occurred on the seasonally inundated grasslands that penetrate far into the *miombo* woodlands of eastern Angola and adjoining western Zambia, along the numerous tributaries of the upper Zambezi River. In the 1960s, perhaps 10,000 wildebeest survived in the south. The largest number (about 6,000) occurred in the southeast, along the Luiana River in Luiana Partial Reserve; this population's range included the Western Caprivi Strip in adjacent Namibia. No information was available on wildebeest numbers in the upper Zambezi region of the country.

In 1975, shortly after attaining independence, Angola descended into a prolonged civil war that did not end until 2002. The country was left in ruins, including its wildlife sector. Very little information is available on the current status of wildlife, apart from the fact that most populations of larger wildlife species were destroyed during the war both inside and outside former protected areas, including the Luiana Reserve.

Zambia

The distribution of *C. t. taurinus* in Zambia is confined to the western region, where there are extensive watershed and floodplain grasslands, notably in Barotseland (Western Province) along the upper Zambezi River and its tributaries and farther east on the Kafue Flats and the Busanga Plain. In the 1960s, little was known about the fauna of the remote western region of Zambia, but it subsequently emerged that

Liuwa Plain NP supports a large, migratory wildebeest population, believed to number about 25,000–50,000 head in the 1980s. Wildebeest also occurred in unknown numbers in some other parts of Barotseland, for example, Sioma Ngwezi Game Reserve in the southwest.

The formerly abundant wildlife of the Kafue Flats suffered from heavy legal and illegal offtake and by the 1960s was largely confined to two large European-owned cattle ranches, Lochinvar and Blue Lagoon, which subsequently became national parks. The decline in the Flats' wildebeest population is illustrated by Lochinvar: it supported 3,000 head in 1937, but by 1966, when it first became a wildlife sanctuary, only 360 wildebeest remained.

Part of the Busanga Plain (960 sq. km) was included in the northern section of Kafue NP when it was established in 1950. In the 1960s, there were approximately 3,000 wildebeest in the park, including about 2,000 on the Busanga Plain.

Since the 1960s, the general integrity of Zambia's protected-area system has been threatened by factors such as escalating poaching, livestock encroachment, uncontrolled fires, and lack of effective law enforcement. There has been a general decline in the country's wildlife populations. But the wildebeest survives in substantial numbers in Liuwa Plain NP, where the population was estimated at 23,500 head in a 2004 aerial survey; this population spends only part of the year within the park, migrating westward toward and beyond the Angola border for several months in the early dry season. This may be the largest surviving population of the nominate subspecies; it is also one of the very few remaining migratory populations of *C. t. taurinus* that still have access to its ancestral wet- and dry-season ranges. The private foundation African Parks took over responsibility for the rehabilitation and management of the Liuwa Plain park in 2003. Consideration may be given to extending the park's boundaries to include more of the wildebeest population's total range. Elsewhere in Barotseland, including Sioma Ngwezi NP, the wildebeest, like most other wildlife, is close to extinction.

On the Kafue Flats, the species is now confined to Lochinvar NP where a small remnant of about 200 survives. The status of the Busanga Plain population in Kafue NP is probably better. It was estimated to number 1,750 in the mid-1990s and may currently be at a similar level. Unlike the rest of the Kafue park, the northern section has been reasonably well protected against poaching over the past ten years thanks to a private-sector initiative.

The total population of *C. t. taurinus* in Zambia is currently about 25,000. This is probably considerably less than in the 1960s, but the extent of the reduction is unclear because of the paucity of knowledge of the size of the major population in the Liuwa Plain area at that time.

Mozambique

The nominate subspecies formerly occurred widely in the Bushveld, floodplain and Mopaneveld of the drier parts of Mozambique south of the Zambezi River. We can only speculate on the former numbers of wildebeest in this region, but it was probably at least many tens of thousands. Wildlife remained widespread and abundant until the mid-twentieth century but was then reduced by uncontrolled shooting, including commercial meat hunting, along with tsetse control programs and the spread of agricultural settlement in some parts of the country. By the 1960s, *C. t. taurinus* was apparently reduced to four surviving populations: Gorongosa National Park, where it still occurred in large but unestimated numbers; the Save River area, where some estimates suggested that tens of thousands of wildebeest still occurred; unknown numbers in an area northeast of the Limpopo River, which subsequently became Banhine NP; and a relic population of about 50 on the Tembe Plains in the far south, an area that supported large numbers of wildebeest until they were massacred to feed troops in the 1930s.

Prior to and following independence, Mozambique suffered a long period of guerrilla hostilities and civil war that finally ended in 1992. During this period, protected areas were abandoned by the government and most of the country's remaining wildlife was slaughtered. By the mid- to late 1990s, the wildebeest was almost extinct in Mozambique south of the Zambezi. The only confirmed survivors were about 30 animals near the Kruger NP border in the south. The species has apparently been lost even from Gorongosa NP, which supported an estimated 14,000 wildebeest when it was abandoned in 1980.

The wildebeest's comeback in southern Mozambique has been initiated by the recent translocation of a few hundred individuals from Kruger NP to Limpopo NP, which is the Mozambique component of the newly established Greater Limpopo Transfrontier Park.

The above summary suggests that the total population of blue wildebeest is currently on the order of 130,000, with the largest numbers in Botswana, South Africa, and Zambia. This undoubtedly represents a marked overall reduction since the mid-1960s, when the subspecies

FIGURE 4.2 (TABLE)   ESTIMATED TOTAL NUMBERS OF *C. T. TAURINUS* IN RANGE
STATES AND TRENDS SINCE 1967 AND 1995

Country	Current Total Population	Trend since 1967	Trend since 1995
South Africa	> 40,000	Increase	Increase
Swaziland	500?	Stable?	Stable?
Zimbabwe	> 10,000	Increase	Stable?
Botswana	46,000	Decrease	Stable
Namibia	> 10,000	Decrease	Increase
Angola	?	Decrease	?
Zambia	25,000	Decrease	Decrease
Mozambique	250?	Decrease	?
Total	>131,750	Decrease	Stable/increase

SOURCE: From Estes and East 2009.

may still have numbered as many as 400,000. Actual numbers in the 1960s are unknown, partly because the precise timing of the massive population declines that occurred in Botswana and Namibia is unclear.

Between the 1960s and 2005, there were particularly severe population declines in Botswana, Namibia, Angola, and Mozambique. These declines have been partially countered by the growth of numbers on private land in South Africa, Zimbabwe, Botswana, and Namibia over the past thirty years. There may now be about 50,000 of the subspecies on private land, which represents about 40 percent of its total numbers.

The current, short-term population trend of the subspecies is stable or increasing, both overall and in most individual range states.

Cookson's wildebeest *(C. t. cooksoni)*

Zambia

Cookson's wildebeest forms an isolated population in the Luangwa Valley of eastern Zambia where it occupies the seasonally flooded alluvial floodplain of the Luangwa River. The narrow floodplain is the dry-season concentration area for all of the valley's grazers and other wildlife. In the wet season, the floodplain is underwater, and the wildebeest is forced back into the *mopane* woodland. The availability of suitably open habitat and pasture during the rains may be the main limiting factor on the size of this wildebeest population. Its numbers were estimated to be only 1,000 in the 1930s but had apparently increased to several thousand by the 1960s.

Both the floodplain and *mopane* habitats have been degraded by uncontrolled burning, resulting in greatly reduced carrying capacity for the wildebeest and other grazers dependent on relatively open country. Poaching has also contributed to reduction of the wildebeest population. Much of this subspecies' range lies within protected areas, centered on North Luangwa NP and adjoining game management areas, with smaller numbers in South Luangwa, Luambe, and Lukusuzi NPs. The core population is well protected in North Luangwa, but poaching has become an increasing problem in most of the valley's other protected areas. Total numbers may currently be more or less stable at between 5,000 and 10,000 head.

Malawi

Cookson's wildebeest occasionally wander up from the Luangwa Valley onto the plateau of Zambia's Eastern Province and across the border into Kasungu NP in western Malawi, but the *miombo* woodland of these areas is completely unsuited to the species, and the dispersal has shown no signs of establishing a viable population.

Johnston's or Nyassa wildebeest *(C. t. johnstoni)*

Malawi

The Nyassa wildebeest formerly occurred in southeastern Malawi, to the east of the Shire River and south of Lake Malawi. The alluvial plains of the river and Lakes Chilwa and Chuita may once have supported a sizable population, but this region is now densely settled; the last wildebeest were shot out in the 1920s.

Mozambique

In northern Mozambique, north of the Zambezi, the larger river valleys and the bushveld and grassland of the coastal hinterland probably supported substantial numbers of Nyassa wildebeest in the past. As in southern Mozambique, after about 1950 the wildlife of the north was greatly reduced by excessive meat hunting, in part to feed the workforce of large-scale plantations of coconut palms, sisal, sugar, and tea. By the

1960s, only three populations of this subspecies were known to survive in Mozambique: in Gile Game Reserve in Zambezia Province, near the upper Lugenda River across the Malawi border from Lake Chilwa, and near the confluence of the Lugenda and Rovuma Rivers in the far north. There was no information on the status of these populations.

No current information is available on the fate of the wildebeest populations of Gile Game Reserve (where the species was on the verge of extinction in the late 1970s) or the upper Lugenda. But the population near the Lugenda-Rovuma confluence has persisted in what is now Niassa Game Reserve and its buffer zone; this was one of the few areas of the country where significant wildlife populations survived the prolonged civil war that ended in 1992. There is now a stable population of about 600–900 wildebeest in this area, mainly in the relatively dry southeastern part of the Niassa reserve's buffer zone. Much of the reserve is dominated by *miombo* woodland, which is unsuitable habitat for wildebeest.

Tanzania

Southeastern Tanzania supports the great bulk of surviving Nyassa wildebeest. Within the region's predominant vegetation of *miombo* woodland, the wildebeest occupies areas of acacia savanna and open grassland that occur primarily in the valleys of the major rivers. In the past it probably occurred in suitable areas throughout the southeast, but by the 1960s, it had become largely confined to Selous Game Reserve and Mikumi NP. No estimate of its numbers had been made in the vast Selous reserve, where its population was considered to be "many thousands." There were only a few thousand in the much smaller Mikumi NP.

Selous Game Reserve suffered from heavy poaching during the late 1970s and early to mid-1980s but has subsequently been rehabilitated with bilateral assistance from Germany. Poaching has been effectively suppressed since 1990, and a development program has been implemented for the reserve and surrounding rural communities. A succession of aerial surveys over the period 1989–2002 has shown that the Selous ecosystem supports a stable population of at least 50,000–75,000 Nyassa wildebeest, with about two-thirds of these within the game reserve and the rest on adjoining lands, including the few thousand in Mikumi NP. This is now the second largest surviving population of the species, after the migratory Serengeti population. Two large concentrations of wildebeest occur during the dry season on short-grass plains on the northern and northeastern boundaries of Selous Game Reserve. Smaller groups are found over much of the rest of the reserve. A wildlife corridor is currently being developed to link the southwestern boundary of the Selous reserve with Niassa Game Reserve across the Ruvuma River in adjacent Mozambique.

Eastern white-bearded wildebeest *(C. t. albojubatus)*

Tanzania

The eastern white-bearded wildebeest formerly ranged widely over the open grasslands and acacia savannas of the Masai steppe in northern Tanzania, east of the Gregory Rift Valley. At the end of the nineteenth century, virtually the whole of eastern Masailand was probably a wet-season dispersal area for migratory plains game, which concentrated around permanent sources of water during the dry season. Numbers are unknown, but there were probably at least hundreds of thousands of wildebeest.

Numbers were subsequently greatly reduced by factors such as disease, hunting, and loss of range and dry-season water supplies to the expansion of livestock and settlement. By the 1960s, only about 9,000 wildebeest were estimated to survive in eastern Masailand. The few remaining dry-season concentration areas for wildlife were in the Rift Valley, in Tarangire NP, and to the north of Lake Manyara. The main wet-season dispersal areas had become confined to the Simanjiro Plains east of Tarangire and the Rift Valley east and south of Lake Natron.

During the 1970s, there was a marked increase in wildebeest numbers in eastern Masailand to an estimated 24,000 head. The reasons for this increase are unknown but might be related to reductions in poaching and/or disease. Most of these animals concentrated in Tarangire NP during the dry season and migrated east to the Simanjiro Plains in the wet season. The population remained stable throughout the 1980s and early to mid-1990s but then decreased rapidly to its current level of about 4,000–5,000. This decline was apparently caused by excessive legal and illegal offtake by meat hunters in the main wet-season range on the Simanjiro Plains. While the bushmeat trade poses the most immediate threat to the region's wildlife, the accelerating conversion of the Simanjiro Plains to agricultural settlement and gemstone mining are the greatest long-term threats to eastern Masailand's migratory ungulates.

Small populations of a few hundred or fewer of the eastern white-bearded subspecies also survive in Lake Manyara and Saadani NPs. The latter, on the Tanzanian coast, is outside the species' natural range. Various species of wildlife, both nonindigenous, including wildebeest, and indigenous, were translocated into Saadani prior to its upgrading from a game reserve to a national park in 2003.

Kenya

The eastern white-bearded wildebeest's range extends northward into the Masai country east of the Rift Valley in southern Kenya's Kajiado District, where the habitat is similar to Tanzania's eastern Masailand. The Kenyan population has declined from historical levels for the same reasons as in Tanzania. By the 1960s, the estimated total number of the subspecies in Kenya was about 12,000, in two more or less separate populations. The larger population (approximately 9,000 head) dispersed onto the Athi-Kapiti Plains in western Kajiado in the wet season and concentrated in Nairobi NP during the dry season. The smaller population (a few thousand head) occurred in southern and eastern Kajiado, with dry-season concentration areas around Lake Natron, at Amboseli, and on the Kuku Plains.

As in Tanzania, the Kenyan population of *albojubatus* increased markedly during the 1970s and numbered about 40,000 by the early to mid-1980s, including 20,000 on the Athi-Kapiti Plains and about 8,000 in the Amboseli area. But numbers have subsequently declined, as Kenya's wildlife has been affected adversely by a massive increase in poaching, as well as drought, disease, competition with livestock, and the lack of both a rational wildlife management policy and an effective organization to implement it.

The Kenyan population of the eastern white-bearded wildebeest now numbers in the low thousands, with most of the survivors in the Amboseli area. The Athi-Kapiti population has been decimated as a result of the expansion of settlement in the Kitengela area to the south of Nairobi NP, which has nearly cut off access by migratory wildlife to the permanent water sources in the park. In recent years, the number of wildebeest seen in Nairobi NP in the dry season has fallen from many thousands to only a few hundred.

Total Numbers of *C. t. albojubatus*

The subspecies' total population increased from about 21,000 in the 1960s to more than 60,000 in the 1980s but has subsequently declined to its current level of perhaps 6,000–8,000. It now ranks with Cookson's wildebeest as one of the two least numerous subspecies.

Western white-bearded wildebeest *(C. t. mearnsi)*

This subspecies occupies the savannas west of the Gregory Rift Valley in northwestern Tanzania and southwestern Kenya (fig. 4.3). It has been separated for millennia from the Eastern race, from which it is different enough to be possibly considered a different species.[10] Early in the twentieth century a small, isolated population of *mearnsi* lived in the Rift Valley on an alluvial floodplain east of Lake Naivasha but was shot out shortly after World War I.[11-14] As the main migratory population defines the Serengeti ecosystem, and is the main focus of this book, I consider its history and current status in the next chapter.

The Mara-Loita Plains population

There is a separate population in the Kenya Mara that migrates east and west within Narok District, between the Masai Mara National Reserve area (dry-season range) and the Masai-owned group ranches on the Loita Plains to the east and north of the reserve (wet-season range). These wildebeest numbered as many as 100,000 until the end of World War II but were reduced to no more than 15,000 head in the late 1940s, after the colonial government opened the Mara-Loita region to uncontrolled commercial meat hunting for several years.[15,16] After this population built up again in the 1970s, commercial wheat and maize farming and fencing shrunk its wet-season range and drought refuges, until it numbers only a few thousand today, including a small subpopulation resident in the Masai Mara reserve and Masai group ranches. The Kenya population is also subject to increasingly heavy poaching for meat, mainly outside the reserve. The fate of this population is unknown to the general public because the arrival in the Masai Mara of half a million or more Serengeti wildebeest in July and August more than fills the void. But for the rest of the year, the wildebeest is the keystone species that was.

PAST AND PRESENT STATUS OF *CONNOCHAETES TAURINUS*

Figure 4.4 shows that the migratory Serengeti population currently comprises more than 80 percent of the species' global population. Adding in the 50,000–75,000 Johnston's wildebeest in the Selous reserve, Tanzania accounts for all but 15 percent of the global population. The increase of the Serengeti population since the 1960s has more than compensated for the decreases in other populations, including the loss of

FIGURE 4.3. Gallery of wildebeest of known sex and age, in order: (a) calf of 4 months, (b) yearling male, (c) 1.5 yr. male, (d) 2 yr. male, (e) 3 yr. male, (f) 1.5 yr. female.

FIGURE 4.4 (TABLE)   ESTIMATED TOTAL NUMBERS OF *C. TAURINUS* IN RANGE
STATES AND TRENDS SINCE 1967 AND 1995

Subspecies	Current Total Population	Trend since 1967	Trend since 1995
C. t. taurinus	130,000	Decrease	Stable/increase
C. t. cooksoni	5,000–10,000	Stable	Stable
C. t. johnstoni	50,000–75,000	Stable?	Stable
C. t. albojubatus	6,000–8,000	Decrease	Decrease
C. t. mearnsi	1,300,000	Increase	Stable
Total	+/– 1,500,000		
Total excluding Serengeti migrants	257,000	Decrease	Decrease

several hundred thousand migratory individuals in the Kalahari. Both the blue and the eastern white-bearded subspecies have declined further in the past decade. This has resulted in both short- and long-term declines in the species' total numbers.

While the wildebeest remains one of Africa's most abundant game species at present, these figures highlight both the uniqueness of the migratory Serengeti population and the vulnerability of the species to further adverse developments, such as those that have affected the eastern white-bearded subspecies during the past decade.

The country-by-country survey of the common wildebeest chronicles humankind's destruction of the ruminant that dominated the acacia savanna ecosystems of eastern Africa.[9] Its story is a particularly egregious example of the ways in which humankind has squandered Africa's heritage of large mammals.

The status of African antelopes at the end of the twentieth century is summarized in East,[17] Vrba and Schaller,[18] Mallon and Kingswood,[19] and Chardonnet and Chardonnet.[20] Appendix 4 in East[17] summarizes the status of all the antelopes in each country of sub-Saharan Africa (including buffalo, giraffe, and okapi).

The decline of wild ungulates is the direct and indirect outcome of competition with *Homo sapiens* and with his livestock and agriculture in particular. Between 1500 and 1900, Africa's human population increased very slowly, from an estimated 5 million to 20 million between 1500 and 1900 in eastern Africa.[21] Near the turn of the century, following the creation of separate colonial countries, Africa's population began to increase exponentially (3 percent per year or more). According to the UN World Population Division's 2008 projection, Africa's population passed one billion in 2010! Although few African mammals have gone extinct in recent times, nearly all are in decline, the trend of most populations is downward, and many populations have been extirpated in developed areas.

The establishment of colonial governments in the nineteenth century set the stage for the population explosion by curtailing tribal warfare, improving public health, promoting commercial agriculture, and adopting policies to increase livestock production at the expense of the native ungulates. The Berlin Treaty of 1884 divvied up most of Africa among Britain, France, Germany, Belgium, Italy, Portugal, and Spain. It required the colonial powers to physically occupy whatever land they claimed with troops, missionaries, merchants, and infrastructure. The artificial boundaries of these newly created states divided tribes and clans, fragmented ecosystems, and cut across wildlife migration routes.

Colonial and postcolonial governments have synergized and greatly increased human environmental impact in the ways exemplified by the history of the wildebeest's decline. Policies designed to increase agricultural and livestock production were pursued regardless of their impact on wildlife in the designated areas. Government efforts to protect cattle from wildlife diseases, implemented by veterinary services, included extermination of wild ungulates thought to harbor such diseases as bovine sleeping sickness (nagana), aerial spraying with insecticides to make tsetse-infested habitats available for cattle ranching (e.g., Botswana, Okavango Swamp), livestock vaccination programs, and fencing.

Fences were the most disruptive, preventing dispersal when erected along international and park boundaries, truncating the ranges of migratory wildlife. The most numerous and destructive were the cordon fences erected in Botswana, largely for the benefit of politically powerful commercial cattle ranchers.[9]

Supposedly installed to separate wildlife from cattle, the fences mainly served to open areas for cattle ranching that had previously belonged to wild herbivores, numbering at least half a million in Botswana and second only to the Serengeti ecosystem. In 1988, blocking the access of central Kalahari wildlife to water resulted in an estimated 50,000 wildebeest dying on the Kuke fence over a period of four months. The World Bank and European Community (EU) have been complicit in the process and not only in-country. By subsidizing the price of beef from South Africa at nearly double the market price in Europe, the EU fueled the drive to export Botswana beef through South Africa.

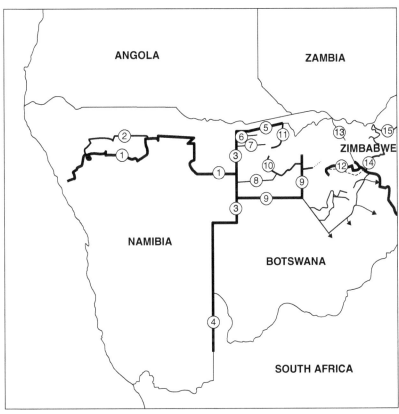

LEGEND

━━━ Major Cordon Sanitaire
━━━ Major buffalo-proof fence
─── Secondary buffalo-proof fence
········ Fence in disrepair
------- Former fence location
──▶ Fence continues

MAP 4.1. Veterinary and international boundary fences erected in northern Botswana and Namibia, neighboring Zimbabwe, from 1958 to 2000. Redrawn from Martin 2005.

Even before exponential growth of the human population began in the twentieth century, proliferation of firearms and commercial exploitation (hides, meat, ivory) had already reduced populations of most big game to very low levels.[22] But the most devastating import of all was rinderpest, the cattle-borne disease that swept from Ethiopia to South Africa in the 1890s, causing both domestic and wild ruminant populations to plummet and recover only many decades later.

The mortality accompanying the severe droughts that visit Africa every decade or so (most recently in 2010) are also devastating to both domestic and wild ungulates. Under such extreme conditions, access to refuges where water and forage can be found enable many migrants to escape starvation and dehydration. Unfortunately for most migratory populations, these refuges are generally in uplands with higher rainfall and more fertile soils—areas with growing human populations. For instance, before the Kikuyu Highlands became densely settled, the Athi-Kapiti Plains migration used to range 50 miles farther north to Fort Hall and the Tana River during severe droughts. Deprived of this drought refuge, their only source of permanent water during the long dry seasons was the Athi River in and around Nairobi NP. They aggregated there in the thousands every year. But access was only possible through the so-called Kitengela corridor in the Masai Group ranches. For decades, conservationists urged the Kenya wildlife authorities to protect the corridor, but all in vain. Meanwhile, the number of registered landowners in the Kitengela area increased from 260 in 1979 to over 20,000 now, almost completely cutting off access to the Nairobi park for the few thousand migratory wildlife of the Athi Plains.[23,24]

Fate of the Eastern White-Bearded Wildebeest in Tanzania: A Case in Point

Unpublished excerpt from original 1967 wildebeest survey, based on Lamprey.[25,26]

In precolonial days, the open grassland and acacia savanna country over virtually the whole of Eastern Masailand formed the wet-season dispersal area for wildebeest, zebra, and Thomson's gazelle, the three most numerous and most migratory plains species. In the dry season, these and other water-dependent herbivores (buffalo, elephant, rhino, waterbuck, warthog, etc., and cattle) concentrated around sources of permanent water, notably near streams and springs at the foot of the western Rift wall, Kilimanjaro, Meru, and several Rift Valley insel-bergs, leaving vast waterless spaces sparsely tenanted by Grant's gazelle, oryx, dik-dik, steinbok, gerenuk, and wandering herds of eland.

The quantity of game that inhabited Eastern Masailand in former times is not known. Until the 1950s there was enough so that appar-ently it never occurred to anyone to attempt a count. Considering that the Serengeti ecosystem currently carries over two million large mam-mals, the far larger area of Eastern Masailand could possibly have sup-ported at least half as many down to the end of the nineteenth century. By then, the destruction of big game by European hunters and Africans newly armed with rifles had already reached such proportions that an international conference of African colonial powers was convened in London in 1899 to consider measures for preserving African animals, birds, and fishes.

Yet it was not outright destruction, or even the catastrophic Euro-pean-introduced rinderpest epidemic that decimated East African her-bivores in 1896, that brought the more prolific plains game of Eastern Masailand to its present low estate. Wildlife had actually suffered less from poaching over much of Masailand than elsewhere, since the Masai lived on their cattle and did not hunt game for meat. Even though cattle contract malignant catarrh from wildebeest at calving time, it never occurred to the Masai (until recently) that the solution was to extermi-nate the wildebeest; they simply kept their cattle away from them dur-ing the calving season. What has dealt the decisive blow to the wildlife of Eastern Masailand is the progressive destruction of their habitat and the loss of their dry-season range.

Hugh Lamprey, who spent four years doing research for his Oxford DPhil on the ungulates of Eastern Masailand, based in Tarangire National Park, documented the loss of most of Masailand's major dry-season water sources:[26]

> Resident farmers on the northwestern slopes of Mount Meru have described to me how, up to 1930, large herds of wildebeest, zebra and eland concen-trated on the land adjoining their farms each dry season. The permanent water which made this concentration possible was that of the small streams which flowed continuously down that side of the mountain and out onto the plains. The water from these streams has since been diverted for the irriga-tion of land and for the watering of cattle at higher levels and this particular dry-season area is now no longer available to the game animals because of the lack of water.
>
> Two mountains adjoining the Rift Valley near Lake Natron, Gelai and Kitumbeini, have small perennial streams which are now entirely exploited

for the domestic animals of the Masai, as also is the only water available in the dry season in the Ngare Naibor river and the Longido springs near the Kenya-Tanganyika border.

Several permanent rivers flow from the south side of Mount Meru towards the Masai Steppe. They flow through farmland and in every case so much water is taken off them before they reach the uncultivated country that none is available for the game animals in the dry season. The largest of these rivers, the Kikuletma, which flows from Mount Meru eastwards to join the Ruvu river, ceased to flow in the dry season in 1952 when the last of the dry season water was removed for agricultural purposes.

These instances illustrate a widespread trend in the reduction in the amount of water available to game animals in the dry season throughout the Masai area of Tanganyika. The extent and distribution of permanent water is probably the most important limiting factor in the numbers and distribution of game animals in the savanna of East Africa and the carrying capacity of the country as a whole is closely related to the carrying capacity of the land within 'cruising range' of the dry season water supplies. (30–31)

Lamprey concluded, "It is probable that the marked reduction in the animals which has been noted in the past 50 years in the Masailand area is associated with the lack of dry-season concentration areas with permanent water" (31). As contributory factors, he points to hunting in the vicinity of European farms around Meru and Kilimanjaro and to grazing competition with domestic stock, particularly in the vicinity of the remaining dry-season water supplies.

Only three important dry-season concentration areas are still available in Eastern Masailand, all in the Rift Valley: two in the Tarangire National Park and the north end of Lake Manyara, where a river from the Crater Highlands flows down the Rift wall and into the lake. "It is thus apparent," wrote Lamprey, "that the carrying capacity of these small areas in the dry season determines the carrying capacity of the Masai steppe as a whole" (32).

Essentially the same situation exists in 2013, except that mining and unsustainable agriculture are steadily shrinking the portion of the Simanjiro Plains available to wildlife.

REFERENCES

1. Hitchins 1968.
2. Attwell 1977.
3. Wilson 1968.
4. Anonymous.[9]
5. Kingdon 1997.

6. Ledger 1964b.
7. Sachs 1967.
8. Estes 1969.
9. Estes and East 2009.
10. Georgiadis 1995.
11. Thomson 1885.
12. Heller 1913.
13. Roosevelt Heller 1914.
14. Meinertzhagen 1957.
15. Simon 1963.
16. Darling 1960a.
17. East 1999.
18. Vrba and Schaller 2000.
19. Mallon and Kingswood 2001.
20. Chardonnet and Chardonnet 2004.
21. Cumming 1999.
22. MacKenzie 1988.
23. Cowie 2004.
24. Cowie 2005.
25. Lamprey 1963.
26. Lamprey 1964.

CHAPTER 5

Increase and Protection of the Serengeti Wildebeest Population

A BRIEF HISTORY OF THE SERENGETI ECOSYSTEM

The Serengeti ecosystem lies on the high interior plateau of East Africa west of the Gregory Rift Valley. From the heights of the Crater Highlands (3,000 m), the land slopes down to Lake Victoria, at 920 m. The Serengeti plains are situated at an elevation between 1,600 and 1,800 m above sea level, a relatively cool island with daily minimum temperature of 15°C, surrounded by hot and humid lowlands; daily minima are 30°C at the lake.

The range of the western white-bearded wildebeest defines the Serengeti ecosystem (map 5.1). It is bounded on the east by the western wall of the Gregory Rift Valley, on the west by Lake Victoria, on the north in Kenya by the Isiria Escarpment and the Loita Hills, and in the south by the transition of the *Acacia-Commiphora* arid savanna of the Masai Steppe to the *miombo* woodland.[2,[1,2,3]]

How, where, and why the wildebeest migrate seems simple enough. They circulate within the range outlined in map 5.1, tracking the greenest, most nutritious pastures. (Movements of zebra and Thomson's gazelle, the other two important migratory species, are considered in chapter 9.) Given the seasonal rainfall pattern of the equatorial climate, with the main dry season lasting many months, migrants move between wet- and dry-season pastures. As there is a rainfall gradient that increases from southeast to northwest, the animals spend the rainy

MAP 5.1. Geomorphic, structural, and volcanic setting of the Serengeti-Mara ecosystem, comprising the physical boundaries of the western white-bearded wildebeest's range. From A. R. E. Sinclair et al., eds., *Serengeti III*, fig. 3.1 in chapter 3 by Peters et al. Chicago: University of Chicago Press, 2008. © 2008 by University of Chicago.

The Serengeti NP (Tanzania) is denoted by a dot-dash line, the Masai Mara National Reserve (Kenya) by a dotted line. These park and reserve boundaries, the Kenya/Tanzania border, and Lake Victoria are based on the USGS *Digital Atlas of Africa* (Hearn and USGS 2001). Numbered volcanoes of the Crater Highlands and Rift Valley are (1) Sadiman; (2) Lemagrut; (3) Oldeani; (4) Ngorongoro; (5) Olmoti; (6) Embagai; (7) Kerimasi; (8) Oldoinyo Lengai; (9) Mosonik; (10) Oldoinyo Sambu. The Rift Valley lakes of Eyasi, Manyara, Natron, and Magadi are shown for reference. Paleo Lake Olduvai, as depicted, was present ca. 1.92–1.70 ma (Hay and Kyser 2001). Rift Valley lakes, main river courses, the Mau Highlands (area above 9,000 ft. [2,744 m]), and the locations of volcanoes are based on the 1982 Tactical Pilotage Chart M-5A (1:500,000 scale). The Mau, Siria, and Kenyan Nguruman Escarpments are based on the 1987 Petroleum Exploration Promotion Project Geological Map of Kenya with Structural Contours (1:1,000,000 scale). The Utimbara Escarpment is based on Barth's (1990) Provisional Geological Map of the Lake Victoria Goldfields, Tanzania (1:500,000 scale). The Natron, Manyara, and Eyasi Escarpments, plus the volcano representations, are based on Dawson 1992, fig. 1. Fault hachures are on the downthrown side of faults. Representation of hills is based in part on Talbot and Talbot 1963, fig. 2. Mara River details are based on 1:250,000 scale topographic maps: Kisumu (Y503, SA-36-4, 1-GSGS) for the uppermost (NE) sections, Musoma (Y503, SA-37-7, I-TSO) for the lower (western) section and the Mara River wetlands.

season in the southeast on the Serengeti short-grass plains. They head west, northwest, and north into the woodland zone when the rains end, most ending up in the Masai Mara late in the dry season. When the rains begin in October–November, the migrants return to the short-grass plains. The distance between these two locations is only 300 km, though movements in between are essentially nomadic and cover a much longer distance.[4,5]

In reality, the system is anything but simple. Researchers have spent over sixty years studying the Serengeti ecosystem. Four volumes of papers exclusively on the Serengeti have been published—*Serengeti I* in 1979, *Serengeti II* in 1995, *Serengeti III* in 2008, and *Serengeti IV* in 2013—not to mention scores of other papers; and still there is no end in sight. The results are like pieces of a great jigsaw puzzle, filling in more details and adding to the big picture of how the ecosystem evolved and functions. Will all the blanks be filled in eventually? Almost certainly not; studies of the insects (notably termites), not to mention the multitudinous soil microorganisms, have barely begun.

Given space limitations, the equivalent of an executive summary is about all that can be offered here. To understand what makes the Serengeti ecosystem such a mecca for herbivores and carnivores, it is necessary to track the geologic history of the region over the past 4–6 million years.[6] The most outstanding landscape features shown in map 5.1 were created by volcanic eruptions and faulting.

Crater Highlands

Tectonic activity leading to major faulting and volcanism over the past 4–5 million years created the main morphological features of the Serengeti-Mara ecosystem.

The Crater Highlands rose in two stages. In Phase I, very fluid lava flowed from a fault running north-south just west of the present Crater Highlands. This so-called flood basalt extended all the way to Kilimanjaro and underlies the Crater Highlands. An exposed part of this scarp forms the west wall of Lake Eyasi at the eastern foot of the Highlands. The top of this deep layer of black rock, dated at 1.9 ma, is exposed at the base of Bed I in Olduvai Gorge.

The seven volcanic mountains of the Crater Highlands rose up along the southeastern border of the Serengeti Plain in Phase II, between 2.5 and 1.7 ma (fig. 5.1).[7,8] The biggest, Ngorongoro, may have been as high as 4,500 m (15,000 ft.); it remained active until about 2 ma, when it collapsed and telescoped into its base, producing the huge caldera with a rim ranging from about 1,400–2,100 m. The Olmoti and Empakaii cones also collapsed. This volcanism occurred during the formation of the Gregory Rift Valley along the eastern edge of the Crater Highlands, as the African tectonic plates spread apart, withdrawing the underlying magma. The Great Rift Valley, of which the Gregory Rift is a part, runs from northern Syria to central Mozambique. The west wall of the Gregory Rift Valley in Tanzania forms the eastern border of the Crater Highlands.

Serengeti Plains

The Serengeti Plains Land Region (map 5.1) is an old peneplain on very old, crystalline rocks that is characterized by a blanket cover of volcanic ash. This forms the landscape type of a generally treeless, gently rolling grassland plain.[9]

Volcanic activity in the Crater Highlands has continued intermittently down to the present, as Oldonyo Lengai, which developed later just east of the Highlands in the Rift Valley, is still active. Before Phase II, the eastern Serengeti Plain was well-watered wooded savanna and grassland similar to the northern extension of Serengeti National Park. Afterward, the prevailing winds from the Indian Ocean dropped abundant rain on the eastern side of the Crater Highlands, leaving the plains downwind in the Highlands rain shadow. Moreover, volcanic ash spewed by periodic volcanic eruptions of the volcanoes and blown downwind contributed to the formation of the eastern Serengeti Plain through the late Pliocene and Pleistocene. Eruptions of Kerimasi (ca. 0.78–0.50 ma) and Oldoinyo Lengai (ca. 0.5 ma–present) deposited vast quantities of volcanic ash in the Olduvai Basin and on the eastern Serengeti Plain (map 5.1). The eastern plains became the driest part of the ecosystem, a shallow, alkaline, very fine sandy loam underlain by calcareous hardpan.[6,10]

Olduvai Basin, at the foot of the Highlands, is the mouth of the river that flows from Lake Makat at Ndutu after heavy rains. It was an alluvial plain where detritus from both Ngorongoro and basement rocks to the west was deposited. About 1.92 ma, a shallow lake formed, which having no outlet was saline and alkaline. Many of the diverse fossils the Leakeys collected and named from the exposed beds along the Olduvai Gorge depended on the 15–20 km diameter lake while it lasted; faulting made it disappear ca. 1.2 ma, early in the deposition of Bed III.[11]

The southernmost Serengeti Plain, south of Lake Ndutu, is predominantly short-to-medium grassland on black-cotton soils derived from calcareous tuff (consolidated volcanic ash).[10]

Lake Victoria, the Mau Upland, and the Mara River

Sediments collected from Lake Victoria indicate that it formed only about 400,000 years ago, in the late Pleistocene (see geologic epochs in table 1.1). It had no outlet until 12,500 years before the present (BP), when it began to flow down the Nile. However, the connection was interrupted for about a millennium around 10,000 BP after the lake level fell below the outlet.

The Mau Escarpment also originated in the late Pleistocene, within the past half-million years. The uplifted and back-tilted upland formed the upper headwaters of the Mara River, the only perennial river in the ecosystem. The Mara River Basin, shared by Kenya and Tanzania, is linked hydrologically with the lake and part of the larger Nile Basin, which is shared by nine countries. The Isiria and Utimbara Escarpments, composed of ancient granitic rock, determine the direction of the river on its course to the lake through Kenya and Tanzania, respectively. An extensive marsh of 100 sq. km lies at the foot of the Utimbara scarp. This marsh and Speke Gulf would have been important parts of the Serengeti ecosystem, especially as drought refuges, before expanding human populations made them inaccessible.

HISTORY OF THE MIGRATORY POPULATION

How has the migratory Serengeti population of the western white-bearded wildebeest escaped the fate of the eastern white-bearded subspecies, and of all the other migratory wildebeest populations reviewed in chapter 4?

A combination of circumstances makes the Serengeti arguably Africa's greatest savanna ecosystem of all time. The western subspecies range lies mainly in Western Masailand (see map 5.1). Masai were still pursuing their pastoral way of life and tolerated wild herbivores. These fierce Nilo-Hamitic warriors had conquered other tribes within the previous three centuries and were still keeping Bantu agriculturalists from settling and farming the rangelands.

The impact of colonial governance and European settlement was far less in Western than in Eastern and Southern Masailand, and still less in

Tanzania than in Kenya. England administered Tanganyika as a UN Trust Territory from 1920 until independence in 1962. There was little or no commercial agriculture, irrigation schemes, fencing, or tsetse control to eat up wildlife habitat and block access to drought refuges. There was also much less overstocking of Masai livestock than in Eastern Masailand. Consequently, the Serengeti ecosystem had remained pretty much intact.

In the 1890s and early 1900s a triple disaster struck. The Italian army imported Indian cattle to the port of Massawa in what is now Eritrea in 1889.[12] Some were infected with the deadly rinderpest virus. They triggered the great rinderpest pandemic that swept from Ethiopia to the Cape Province of South Africa in seven years. The virus killed 95 percent of the cattle in East Africa within two years, along with wildebeest and buffalo, the most susceptible and abundant of the wild ruminants, as well as eland and other spiral-horned antelopes, giraffe, roan, bushpig, and warthog. The Masai, dependent on a diet of milk and blood, proceeded to starve; many also fell victim to an epidemic of smallpox.[1,13,14] At least two-thirds of the whole tribe died. Neighboring Sukuma and Kiria tribes, pastoralists who also farmed, fared only slightly better. Cattle-rustling Masai warriors pushed them back from their common eastern boundary and thereby furthered the spread of rinderpest. Abandonment of villages and the reduced ability of survivors to grow food led to repeated famines from 1900 until 1920.

Lions deprived of wild and domestic herbivores became maneaters. They scourged the country for a quarter century, further hastening the abandonment of human settlements.[1]

Depopulation of the herbivore and human populations enabled woody plants to replace the savanna grasslands and abandoned cultivation plots with woodland and dense thickets, the preferred habitat of tsetse flies (*Glossina* spp.). By 1910, many of the wild ruminants had developed immunity to rinderpest and had again become numerous. Tsetse flies feasted on their blood and infested the woodlands. Protozoans that cause bovine sleeping sickness (nagana) and different forms of human sleeping sickness (trypanosomiasis) are transmitted by tsetses. The game had immunity to nagana, but cattle couldn't survive in infested areas. Periodic outbreaks of human sleeping sickness in western Serengeti speeded the abandonment of agricultural villages.

Rinderpest outbreaks in 1917–18, 1923, and 1938–41 continued to affect wild ruminants and soon became endemic in the cattle populations surrounding the Serengeti.[1]

In Kenya's part of the Serengeti ecosystem, too, drought followed by the rinderpest pandemic in the 1890s decimated cattle, wildebeest, and buffalo, followed by famine and disease. Removal of large herbivores and humans enabled woody vegetation to take over the plains; by the 1930s and up to the 1950s, woodland replaced the Masai Mara grasslands.[15,16]

HAPPY HUNTING GROUND

The remarkable recovery of the wildlife and the near-absence of people at the end of the nineteenth century and for many years in the twentieth made the Serengeti a hunter's paradise. It was already famous for its lions as early as 1910.[1] Safaris of famous people led by famous white hunters were reported in the international press and in their books[17–20] and films.[21,22]

Most safaris came from Kenya via Narok, across the Masai Mara into Tanganyika, and as far as what is now Seronera. This road still exists, although the border has been closed to the public since 1977. En route, Percival[20] in July 1912 and White[23] in August 1913 encountered huge dry-season concentrations of wildebeest along the Mara River. No one at the time knew where they came from. It remained for the films Martin and Osa Johnson took during their 1926, 1928, and 1933 expeditions to alert the world to the great wildebeest migration, adding to the fame of the Serengeti ecosystem.

Unrestricted and unrestrained hunting in the Serengeti and generally in British East Africa led to such excessive killing that Sir H. H. Johnston, the renowned Africanist, wrote in the foreword to Schillings's (1905) *Flashlights in the Jungle,* "If any naturalist explorer previously deprecated the frightful devastation which followed in the track of British sportsmen, and a few American, Russian, German, or Hungarian imitators, it was thought that he did so because he was a bad shot, or lacked the necessary courage to fire at a dangerous beast."

In 1921, a year after the League of Nations gave Britain the mandate to govern Tanganyika as a UN Trust Territory, the new government introduced the first game laws.[24,25] Although the Germans had built a fort at Ikoma, they had not established an administrative apparatus in the area.

It was not until 1928 that the British administration got around to gazetting a Serengeti game reserve. Ngorongoro Crater was established as a Complete Reserve bounded by the Crater rim. In the early 1930s, all the country from the Rift Valley to Lake Victoria and from the Kenya border to the southern boundary of the Serengeti, including Ngorongoro

Crater and the surrounding Highlands, comprising the Loliondo sub-district, was placed under government control. Permission of the provincial commissioner was required for anyone to enter the area for the purpose of hunting or photography. But those who were permitted to enter what was then known as the Serengeti Closed Reserve, the section that ran from Seronera to Lake Victoria, were free to hunt without limits or fees until 1935.

Outrage over the shooting of "tame" Banagi and Seronera lions, accustomed to having kills dragged to them behind safari vehicles, caused shooting of lions to be prohibited in the neighborhood of these two camps. Henceforth, only one lion could be hunted elsewhere in the reserve on a Resident's or Visitor's Full Game License. In 1937, the authorities placed a complete ban on hunting lion, leopard, cheetah, hyena, wild dog, rhinoceros, buffalo, and roan in the whole Serengeti Closed Reserve.

In 1940, the Tanganyika government enacted a game ordinance, under which areas could be declared national parks. The purpose was to implement the provisions of the 1933 Convention Relative to the Preservation of Fauna and Flora in Their Natural State, of which the Tanganyika government was a signatory.[25] The Serengeti Closed Reserve was chosen to become Tanganyika's first national park, with the same boundaries.

The Masai, for whom grasslands inside the park provided important forage, had not been informed of these decisions.[24] As national parks by definition excluded people, apart from managers and previously established residents, some ten thousand pastoralists were suddenly to be denied entrance without a permit. Adding insult to injury, it was made unlawful to burn the grassland or bush within the national park, as was the Masai practice from time immemorial. But no action was actually taken under this legislation, and the Masai continued to move with their cattle in and out of the designated park until it was replaced by the National Parks Ordinance in 1948, which repealed many sections of the 1940 Game Ordinance.[24]

A board of trustees, charged with running the nation's parks, replaced the Game Department, which the 1940 Game Ordinance had placed in control. Still, this new ordinance remained unenforced until 1951, when a proclamation was published defining the Serengeti National Park boundaries as the southern Serengeti and the Ngorongoro highlands. In 1952 park headquarters were established at Ngorongoro, with a western outpost at Banagi, then the park's northern boundary.

Next, the district commissioner told the Ngorongoro Masai what was in store for them. Their rights would be safeguarded but only applied to those who had been living in the park when it was proclaimed in June 1951. No additional Masai would be allowed to move in, and no more livestock than were present at that time were allowed. If necessary, the numbers of livestock could be reduced should their population exceed the carrying capacity of the range. One can just imagine how the Masai, for whom wealth was measured by the number of cattle owned, reacted to this restriction.

Finally, no cultivation was to be allowed in the park. All Arusha, Mbulu, and other cultivators then present would have to leave. The possibility that the Masai might in future want to supplement their diet with gardens, or were presently married to Arusha cultivators, was ignored. As of 1954, there were fifty cultivating households in Ngorongoro Crater and another fifty households in Empakaai Crater.[24] Eventually, with government assistance, the cultivators were resettled outside the park, and that problem was solved—temporarily.

Resistance and outright refusal of the Masai to accept the new park boundaries and some of the restrictions led to consultations between the government and the park board of trustees, which unexpectedly sided with the Masai. The board recommended breaking up the Serengeti into three smaller parks: the western corridor running to Lake Victoria, Ngorongoro Crater together with the Northern Highlands Forest Reserve, and Empakaai Crater. Very little of the true Serengeti short-grass plains was included. The bottom line: the Masai's needs were given first priority over conservation of the wildlife.

Publication of these proposals in the notorious White Paper (Sessional Paper No. 1 of 1956)[26] created such an international furor that the government decided to appoint a three-man committee of enquiry to consider the flood of opposing proposals made by international and governmental organizations. A 1956 film, *No Room for Wild Animals*, by Michael Grzimek,[28] posthumous coauthor with his father, Bernhard, of *Serengeti Shall Not Die*,[27] protested the British Government of Tanganyika's decision to cut the area of the park by one-third. The film was particularly effective in arousing international concern. After winning prizes at a Berlin film festival, it was shown in Britain (London), the United States (New York), and sixty-one other countries.[28]

Following consultation with the British Colonial Office by the Tanganyika government, W. H. Pearsall, professor of botany at London University, was sent to make a scientific evaluation of the Serengeti situation on behalf of the Fauna Preservation Society. His report,[29] based on a two-month visit to the area in late 1956, had the greatest influence and was accepted as the scientific basis for the decision of the Committee of Enquiry when it met in 1957. However, not all of his recommendations were adopted. Most notably, the committee's recommendation to extend the White Paper's proposed park eastward to incorporate the full extent of the short-grass plains was rejected because of Masai claims to the area, regardless of its critical importance to the wildebeest migration.

The truth was that no one at the time understood much about the migration. Pearsall and others who knew the Serengeti best believed the migration went east and west and included the Ngorongoro Crater. Two reports and one book-length study by the Grzimeks documenting the main migration routes in 1959–60 showed that the migrants used the eastern plains but not the Ngorongoro highlands or the Crater.[27,30,31] The root of the controversy was whether traditional land rights could be allowed in national parks.

The committee finally decided that human rights that conflicted with wildlife conservation could not be allowed in a national park. But in exchange for giving up their right to go west of a line demarcating the eastern park boundary, the Masai were granted the right to occupy the Crater Highlands. The committee's report proposed setting up a conservation unit to administer this area. The idea was accepted by the government and even expanded to include more of the highlands and much of the short-grass plains. Ngorongoro and Empakaai craters were to receive special status as Nature Sanctuaries.

In 1959, Ngorongoro was excised from the national park and established as the Ngorongoro Conservation Area. Park headquarters were moved to Seronera. The NCA became the first multipurpose protected area in Africa, wherein the needs of people and wildlife were considered equal. Fortunately, the realignment of the park boundaries entailed the addition of the area north of Banagi to the Kenya border, thus providing a link with the Masai Mara, which Talbot and Talbot[32] showed in 1961 was the essential route to the dry-season refuge on the Mara River for the Serengeti migration.

An area of 647 sq. km (250 sq. mi.) enclosed by the Mara River, the Isiria Escarpment, and the Kenya-Tanganyika border was gazetted in 1948 as the Mara National Reserve. It was known as the Mara Triangle. But in 1961 Kenya's national reserves were all degazetted. Management of the renamed Mara Game Reserve, enlarged to 1,800 sq. km (700 sq. mi.), was turned over to the Narok District Council, with a Masai game warden in charge.

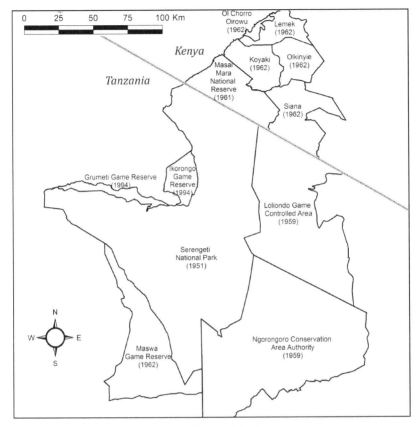

MAP 5.2. The protected and unprotected areas of the Serengeti ecosytem and dates of gazetting. Courtesy of Grant Hopcraft.

Later additions to the protected parts of the Serengeti ecosystem and their dates of gazetting are shown in map 5.2. As you can see, most but not all of the wildebeest's range is included.

EXPONENTIAL GROWTH OF THE SERENGETI POPULATION

When I began my wildebeest research in 1963, the Serengeti population numbered no more than a quarter million animals according to the first complete census, conducted in 1961[32,33]. Viewed from Naabi Hill, the entrance to Serengeti National Park from the Ngorongoro Conservation Area, the spectacle of many thousands of wildebeest dotting the surrounding short-grass plains was just as awesome in 1964 as it is today. Since then, the population has increased to an estimated 1.3 million animals, but only so many can be viewed from one vantage point.

The increase began in 1962, after rinderpest finally lost its hold on the wildebeest. Previously, in recovering from the rinderpest pandemic that destroyed 90 percent of the Serengeti population in the 1890s, adults developed antibodies that conferred immunity to the disease. But calves lost the colostrum immunity derived from their mothers' milk by seven months. By the end of the dry season, the nutritional low point, rinderpest had increased the calf mortality rate by some 10 percent. Assuming the annual calf crop equals 30 percent of the population, by the end of the year only 8 percent survived to the yearling stage.[2] The late dry-season die-off of calves infected with rinderpest was first detected in 1933 and became known as the "yearling disease."[32]

Although cattle also developed immunity as adults, rinderpest was endemic in the area surrounding the Serengeti wildebeest and the source of the yearling disease. The East African Veterinary Services had begun inoculating cattle in the Serengeti region to immunize them against rinderpest in the 1950s. Suddenly, in 1961, rinderpest no longer infected cattle. Within a year, it also disappeared from the wildebeest, as shown by shooting and taking blood samples from hundreds of animals in Serengeti NP and the NCA.[34] From 1963 on, calves no longer developed antibodies to the disease. From then until the end of the decade, yearling survival rose from 25 to 50 percent, and the wildebeest population increased at the rate of 10 percent a year. Total counts in 1963, 1965,

---

Virus Deadly in Livestock Is No More, U.N. Declares

By DONALD G. McNEIL Jr.
October 15, 2010, New York Times

The last case was seen in Kenya in 2001. On Thursday, the United Nations Food and Agricultural Organization announced that it was dropping its field surveillance efforts because it was convinced that the disease was gone. The official ceremony in which the World Organization for Animal Health will declare the world rinderpest-free is scheduled for May. (That organization, known as the O.I.E. for its initials in French, was created in 1929 chiefly to fight rinderpest.)

1967, and 1971 showed that the population had doubled to half a million. But between 1965 and 1970, there were intervals when calf survival rates declined; it seemed that poor dry-season grass production was checking the population increase.

Beginning in 1971 and continuing through 1976, dry-season rainfall (July–October) in the northern woodlands increased from an average of about 150 mm to 250 mm. The increased grass production that resulted enabled more wildebeest to survive the dry season, and the population began increasing again. Finally, in 1977, wildebeest numbers stabilized at 1.3 million and have since remained around that level, give or take a quarter million.[35]

CONSERVATION OF THE SERENGETI ECOSYSTEM

The United Nations Educational, Scientific and Cultural Organization (UNESCO) inscribed Serengeti National Park as a World Heritage Site and International Biosphere Reserve in 1981. The Ngorongoro Conservation Area was given World Heritage Site and International Biosphere Reserve status in 1979. Thus both have been singled out by UNESCO as "protected areas of global importance, the conservation of which is the concern of all people."

REFERENCES

1. Sinclair 1979a.
2. Sinclair 1979b.
3. Sinclair et al. 2008.
4. Thirgood et al. 2004.
5. Hopcraft 2010.
6. Peters et al. 2008.
7. Pickering 1968.
8. Dawson 1992.
9. Gerresheim 1974.
10. Anderson and Talbot 1965.
11. Hay and Kyser 2001.
12. Reid 2012.
13. Baumann, in Fosbrooke 1963.
14. Sinclair et al. 2007.
15. Dublin et al. 1990.
16. Darling 1960a.
17. Selous 1908.
18. Schillings 1905.
19. Hunter 1952.

20. Percival 1928.
21. Johnson and Johnson 1928.
22. Johnson and Johnson 1934.
23. White 1915.
24. Fosbrooke 1972.
25. Lithgow and van Lawick 2004.
26. Tanganyika Legislative Council 1956.
27. Grzimek and Grzimek 1960a.
28. Grzimek and Grzimek 1956.
29. Pearsall 1957.
30. Grzimek and Grzimek 1960b.
31. Grzimek and Grzimek 1960c.
32. Talbot and Talbot 1963b.
33. Talbot and Stewart 1964.
34. Plowright and McCulloch 1967.
35. Mduma, Sinclair, and Hilborn 1999.

CHAPTER 6

Serengeti Grasslands and the Wildebeest Migration

The gnu's specialization for a particular range of grassland habitats, conveniently identified as acacia savanna, was outlined in chapter 2 and compared in chapter 3 with topi and hartebeest, two members of the same tribe with which it associates in the Serengeti ecosystem. Their numbers and their movements can hardly be compared to the wildebeest's, whose migration covers the entire Serengeti ecosystem (fig. 6.1; see maps 6.1, 7.1).

The straight-line distance from end to end of the wildebeest migration, from the Serengeti plains to the Masai Mara, is about 325 km. So a round trip would cover 650 km. But a sample of wildebeest fitted with satellite collars showed that an individual gnu traveled an average of 4.25 km a day, amounting to 1,550 km in a year (map 6.1).[1,2] In another study involving ten collared wildebeest, both males and females averaged 7 km a day year-round.[3]

Unsurprisingly, grazing preferences and water sources govern these seasonal movements. A model using collared wildebeest to track their movements in both the wet and dry seasons predicted that wildebeest would move onto ranges where grasses were at their most nutritious stage. They could maximize their energy intake on swards 3 cm high and maintain energy balance on swards between 3 and 10 cm high. Preference for short- and intermediate-height grass of moderate greenness during both the wet and dry seasons would largely account for migratory movements.[4] But to understand the migration, the major

FIGURE 6.1. *(Two-Page Spread)* Serengeti migration: a) Wildebeest migration spread out over the short-grass plain, near Naabi Hill; b) Moving off the short-grass plains at the end of the rainy season; c) An army on the move; d) Waiting to drink and cross the river at midday; e) Drinking and crossing the Grumeti River; f) Mixed zebra concentration at Seronera, at the border between the Serengeti plains and the woodland zone; g) Mowing the long grass in the Grumeti Reserve; h) Successful search for short green pasture on the Sabora Plain; i) A wildebeest throughway created by thousands upon thousands of hooves; j) Traffic jam; k) Vanguard of the migration entering the Masai Mara Reserve; l) Crossing the Mara, the one perennial river in the Serengeti ecosystem.

FIGURE 6.1. *Continued*

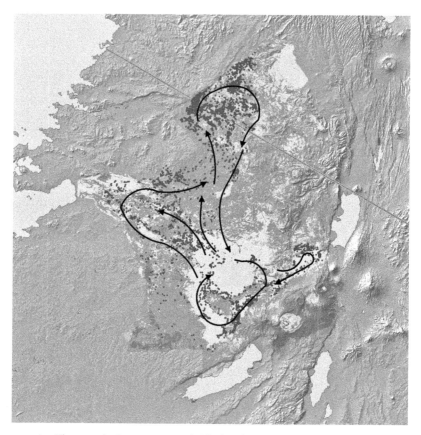

MAP 6.1. The annual migratory route of eight female wildebeest collared between 1999 and 2001 in the Serengeti-Mara ecosystem. Map by J. G. C. Hopcraft, Frankfurt Zoological Society, and the Centre for Research into Ecological and Environmental Modelling, University of St. Andrews.

differences in the Serengeti grasslands, based on geology, rainfall, and soil characteristics, need to be considered more closely.

THE SERENGETI SHORT-GRASS PLAINS

Your first impression of the Serengeti Plain depends on the time of year you see it. If you come in the dry season, June–October, typically overland from Ngorongoro, the vista that opens before you is the tan color of lions. Emerging onto the plains in the Ol Balbal at the foot of the Crater Highlands, you proceed across a dusty, waterless landscape sparsely populated with plains game that are water-independent: giraffe, gazelles, eland, dik-dik, ostrich. You won't see any wildebeest or zebras on the short grasslands. They left in May once the rainwater pans dried up, possibly even before the grass stopped growing.

If you come in the rainy season, you will be overwhelmed by the spectacle of thousands upon thousands of antelopes and zebras spread across unending green pastures. I imagine I'm looking down golf course fairways, one after another, receding into the distance. The migratory species congregate here as soon as the rainy season begins, accompanied by the scavengers and predators that attend them. For most, the season of peak births is between December and March.

Why do all these herbivores come here to stay during the rains? The grass is green throughout the ecosystem, and there's much more of it in the higher rainfall regions to the west and north. Water to drink can be found everywhere. What's so special about the short-grass plains?

Samuel J. McNaughton, professor emeritus of biology at Syracuse University, who spent three decades studying the Serengeti grasslands and herbivores, presented evidence in a letter to the journal *Nature* that the seasonal movements of migratory grazers in the Serengeti ecosystem are related to mineral content in the forage.[5] Based on nutrient concentrations in the grass and on animal nutritional requirements, calcium (Ca), copper (Cu), nitrogen (N), sodium (Na), and zinc (Zn) are particularly important; and for lactating females and growing young, magnesium (Mg), phosphorus (P), and the calcium/phosphorus balance.

The availability of these elements to grasses and their grazers depends on their location in the soil parent material and on the capacity of the soil to adsorb and exchange elements with the vegetation. The ions of the above minerals, called cations, are positively charged; they bind to negatively charged soil particles. The finer the soil, the larger the surface area to which cations can bind. Accordingly, clays have the greatest ability to adsorb and exchange cations with plant roots. This cation exchange capacity (CEC) is highest in soils with a neutral or higher pH and lowest in acidic soils.

The seasonal water balance (salinity) and biotic processes—leaching from plants, decomposition of litter to form humus—are additional factors that determine soil fertility.[6] Furthermore, the herbivores themselves are by far the main source of nitrogen and ammonia (NH_3), through elimination of dung and urine. (More about that later.)

In his *Nature* article, McNaughton[5] showed that the dry- and wet-season ranges of the Serengeti migrants were nutritionally completely distinct. Seasonal usage reflected the mineral content of the preferred forage during the rains. The wet-season range supported grass of consistently higher mineral content. Preferred grasses were high in iron, nitrogen, magnesium, and calcium. Forages from the dry-season range were high in manganese and zinc, while those from transitional areas were high in copper, magnesium, sodium, and potassium. Magnesium, sodium, and phosphorus seem particularly important.[7]

Overall, sodium content was three times higher in the seasonal concentration areas. Aluminum and cation exchange concentrations were 80 percent higher potassium, 50 percent higher manganese, and 40 percent higher lead; and calcium, magnesium, and vanadium were 10–23 percent higher than in areas of nonconcentration.[7]

The presence of grasses that accumulate sodium in their leaves is a distinctive property of the Serengeti grasslands. Grasses with high sodium content are prominent in transitional and wet-season ranges. Grazers that can meet their sodium requirement just by eating grasses that accumulate the mineral have no need to supplement their diet by eating saline deposits (salt licks). Such pastures may influence the areas where migrants concentrate. Likewise, the grasslands where resident grazers concentrated were found to be high-sodium areas.[5]

The grasslands considered in the Serengeti rangeland study occur along a fertility gradient from ancient, infertile granitic soils in the dry-season range to recent, fertile volcanic soils in the wet-season ranges.[8,13] This west-to-east forage-mineral gradient runs opposite to the rainfall gradient.

Clearly, the reason the migration congregates on the short-grass plains is that they offer the most nutritious forage in the Serengeti ecosystem. However, these pastures can only be utilized while green and growing. In the rain shadow of the Crater Highlands, the eastern-to-central Serengeti Plain is the most arid part of the ecosystem.[8]

Repeated eruptions in the Highlands over a span of millions of years have formed a sediment plain overlying the ancient peneplane. Prevailing easterly winds have deposited ash up to 16 m deep at Ndutu, over 45 km from the source. The most recent windblown dust from Oldonyo Lengai can be seen north of Olduvai Gorge as mobile dunes. Other older dunes in this area are vegetated and stabilized.

Lengai is the only known active volcano that spews natrocarbonatite, a highly fluid lava containing almost no silica, in contrast to other volcanoes. Natrocarbonatite, named for nearby Lake Natron, is composed chiefly of sodium, potassium, and calcium carbonates.[9]

Dust and ash from Lengai explosions over the past half-million years (most recently in 2008) have blown great distances westward across the peneplane, the accumulation diminishing with distance from the Crater Highlands. The soil of the short-grass plain is a shallow, alkaline, very fine sandy loam overlying a calcareous hardpan formed by leaching of the lime-rich soil. The plain is largely treeless except along Olduvai Gorge, as the hardpan is impenetrable by deep-rooted plants and grasses except where broken or worn away along drainage lines, rocky outcrops, and waterholes. But also thanks to the hardpan, minerals that would be unavailable in deeper soils nourish the short, colonial, rhizomatous grasses carpeting the plain. Common grasses like dropseed *(Sporobolus)*, finger grass *(Digitaria)*, mat-forming blue grass *(Andropogon)*, and star grass *(Cynodon dactylon)* and the dwarf sedge *Kyllinga* are adapted to the short growing season and heavy grazing.[8,10–13]

As is typical of arid savanna, rainfall is sporadic and unreliable both in occurrence and in distribution. Strong dry-season winds and a hot, dry climate cause considerable areas to support only a sparse grass cover, especially in the roughly half of the short-grass plains within the Ngorongoro Conservation Area (8,288 sq. km), including Olduvai Gorge and the Salei Plain, where desert-adapted oryx dwell (see map 7.1).

Now and again enough rain falls on this arid area to produce abundant green pasture. Then the migration comes all the way to the Ol Balbal at the foot of the Crater Highlands, where they may meet several thousand wildebeest drawn from Ngorongoro Crater when the rains begin sooner outside than inside. The runoff from Lake Masek down the Olduvai Gorge may even form a temporary shallow lake in the Ol Balbal. Thousands of wildebeest cows drop their calves and may stay for several months on these plains, even after the main migration has withdrawn westward. Once the rains end, many of the Serengeti migrants follow the Ngorongoro contingent into the Crater, adding thousands to the dry-season wildebeest count (see chapter 7).[14]

A pause during the rainy season long enough for the grass to stop growing makes wildebeest move off the short-grass plains to transitional areas of higher rainfall. The Maswa Game Reserve (2,200 sq. km) southwest of the park serves as a refuge during the December–April period (see map 7.1). It is the closest woodland area to the southern plains. *Acacia-Commiphora* woodland also surrounds Lakes Ndutu and Masek. South of Ndutu, a largely treeless plain of medium *Pennisetum/Cynodon* grassland growing on black-cotton soil (alluvial clay loams) extends to the edge of the woodland. Wildebeest and other plains game that have deserted the short grasslands can be seen by the thousands along the dirt road to Makao.

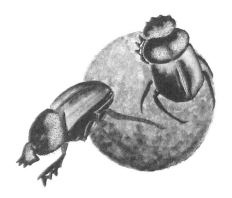

FIGURE 6.2. A pair of dung beetles using celestial navigation to line out the way to a suitable burial ground. Painting by Dino Martins.

THE GARDENERS OF EDEN

The herbivores thronging the short-grass plains are not just consumers. Grazing stimulates new growth and serves to maintain high nutrient–rich forage that increases ungulate carrying capacity.[5,7,15] A concentration of grazers crops all herbaceous plants to a height of 3–5 cm. The animals give back what they eat as dung and urine; and the effect of trampling by millions of hooves could be likened to plowing and harrowing a field. No mechanical manure spreader could cover the ground more efficiently. Thus they sow and can reap again within a few short weeks as luxurious new growth covers these fertilized pastures.

Dung beetles are an integral part of this recycling system (fig. 6.2). This marvelous recycling system is largely limited to the rainy season, when and where the migration is present and the ground is soft enough to dig. In the dry season, there is little or no dung beetle activity, and the

A single dung beetle can bury 250 times its own weight in a night. The beetles lay their eggs in the dung, which provides the food required by the next generation of beetles. Watching the big scarabs making, rolling and burying dung balls is one of the must-see sights at this time of year. Meeting at a dung pile, a male and female pair up. The male offers her a giant dung ball 50 times his own weight and the two roll it away, or he rolls while she rides on top. They find a soft place and bury the ball before mating—mostly underground. She lays a single egg in the ball; females of many of the larger "rollers" stay underground and care for and protect their eggs and young, while the male pursues other mating opportunities. (From Facts about Insects, uploaded from Google)

FIGURE 6.3. *Cynodon dactylon*, a shallow-rooted, nutritious grass that carpets the short-grass plains. From Tallmark, Milbrink, and Backeus 1998 (mimeograph).

ungulates' droppings (pellets of antelopes, cowpats of cattle and buffalo, road apples of zebras and donkeys, boluses of elephants) lie on the surface until the rains return and the fecal particles percolate into the soil.

Naturally, the interaction between the vegetation of the short-grass plains and its consumers is not that simple. One major difference between herbivory of the short-grass plains during the rains and of the long grasslands during the dry season is that grazing the short grasses keeps them growing but reduces seed production by removing most of the flowering stalks. They only get a chance to flower and produce seeds after the migration has moved off the plains, and by then the growing season is almost over. In longer grasslands largely ungrazed during the rains, the grasses flower before defoliation and reproduce through seeds, which sprout at the beginning of the rains (see below). But observations of a short-grass study plot found very few seedlings of perennial species, suggesting that only those able to propagate vegetatively could reproduce successfully.[16] Rangeland studies have shown that removal of seed heads can stimulate vegetative growth. Grasses that put out tillers—stolons aboveground and rhizomes belowground—are able to reproduce very efficiently in the vegetative stage and can spread rapidly over or under the soil. In fact, 58 percent of the perennial grasses in the short-grass study plot were clonal.

Grasses like Bermuda or star grass *(Cynodon dactylon)* (fig. 6.3) and Kikuyu grass *(Pennisetum clandestinum)* are called mat or couch grasses. They spread by putting down roots at the nodes. Vegetative tillers also tend to be less stemmy and more nutritious than reproductive tillers.[17] In this way species of sod grasses and mat grasses compete to occupy the bare spots between clumps. Grasses with a greater capacity to grow following defoliation expose larger leaf areas for photosynthesis and attain a competitive advantage.[18]

Annual grasses avoid loss of their flowering parts by hugging the ground; physically, by having protective hairs, pointed bracts, or awns on their flowering heads to reduce their palatability; chemically, by producing distasteful or indigestible compounds—or by flowering late in the growing season after most of the large herbivores have migrated away (e.g., *Chloris pycnothrix*).

Overgrazing of the *Sporobolus/Kyllinga* short grasslands can greatly reduce the grass cover. Such areas can be seen in the NCA subjected to grazing by both domestic stock and wild animals. Comparative studies indicate that such overuse does not occur on grasslands grazed by wildlife alone.

TRANSITIONAL GRASSLANDS

Driving northwest on the road to Seronera, the nature of the soil and grassland types changes with increasing distance from the shadow of the Crater Highlands, as rainfall increases and evapotranspiration decreases. The soils become more mature and deeper from east to west and vegetation types also change, but without any sharp boundaries. Soon after passing Naabi Hill, the park entrance, you cross a grassland of medium height dominated by *Cynodon* and *Sporobolus* grasses, growing on lime-rich soil underlain by a softer pan that woody roots can penetrate here and there (fig. 6.4). This type is intermediate between the short and long grasslands. It is relatively heavily grazed during the rainy season.[8] The landscape here is dotted with large termite mounds, absent in the short-grass plains, which indicate changes in the plains habitat. Herds of topi and hartebeest prefer this area; coincidentally, they like to stand on the mounds to see and be seen.

The medium grassland gives way to long grassland, 30 cm (12 in.) high or more, dominated by bamboo *(Pennisetum)* and bluestem *(Andropogon)* grasses. The soil is black cotton and still at this distance volcanic in origin. The migrants may graze these grasses on the way to and from the woodland zone, selecting various shorter, more palatable species beneath the overstory of tall grass. This association is only lightly grazed during the rains, probably because when wet the clay sticks to hooves and makes walking difficult. It also makes the herbivores more vulnerable to predators. I have seen hyenas suddenly become

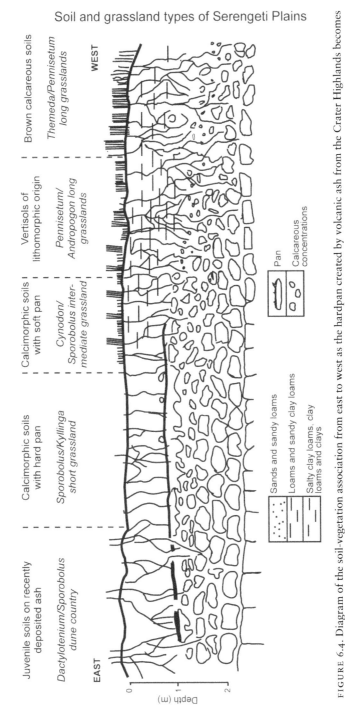

FIGURE 6.4. Diagram of the soil-vegetation association from east to west as the hardpan created by volcanic ash from the Crater Highlands becomes thinner and softer, allowing deeper-rooted grasses and woody plants to grow. From Anderson and Talbot 1965.

active in late morning after a downpour handicapped a concentration of wildebeest foregathered on these pastures.

The last soil/grassland type before entering the woodland zone at Seronera is tall red oat *(Themeda triandra/Pennisetum)* grassland growing on brown calcareous soil. It is lighter textured and better drained than the previous (black-cotton) type. An abundance of palatable understory grasses causes this range to be heavily grazed early and late in the wet season and moderately in the dry season.[8]

DRY-SEASON RANGE IN THE WOODLAND ZONE

The woodlands grow on the mature, less mineral-rich soils of the ancient granite shield. Occasional rocky outcrops called kopjes are among the oldest (Precambrian) rocks on the planet. Increasing rainfall and lower evapotranspiration produce a more vigorous undergrowth. With the ratios of rainfall to evaporation far narrower than on the hotter, drier, short-grass plains, the soils are subjected to rich weathering and weak leaching, resulting in the accumulation of soluble salts, carbonates, and bases.[8] Deeper soils support larger tussock grasses, such as species of *Digitaria, Themeda,* and *Eustachys* on well-drained soil and *Pennisetum* on poorly drained soils.[19]

Since primary production increases with rainfall, migrating herbivores should go to the wet end of the gradient, especially the medium and large bulk feeders like wildebeest, buffalo, and zebra. However, forage quality generally declines with rainfall,[6,20] as the taller grasses are more fibrous and harder to digest. Translocation of nutrients into their roots after flowering reduces nutrients during the dry season, unless growth is stimulated by thunderstorms or burning. Given the high levels of calcium, phosphorus, and potassium in all soils, the herbivores' grazing patterns may depend mainly on the palatability of grasses at preferred stages of growth.[21]

If the migration simply followed the rainfall gradient, the hordes should all trek northwest to remain on green pastures until they reached the Masai Mara. Instead, perhaps half or more of the wildebeest, preceded, accompanied, or followed by the zebras, move into the western corridor (see map 7.1). Armies of them move along the Mbalageti and Grumeti River valleys, which empty into the Speke Gulf salient of Lake Victoria. Other armies mow the medium and tall grasslands as they move west across the extensive plains between the two rivers: Musabi, Nangangwa, Sibora, Kawanga, Dutwa, and Kirawira.

The western part of the corridor, terminating in the Ndabaka Plain, has deep black-cotton soils deposited during the Pleistocene after faulting caused a shoulder in the Western Rift to uplift and dam the west-flowing drainage, creating several lakes in addition to Victoria, the largest. These alluvial deposits make corridor grasslands exceptionally mineral-rich. But they are only accessible in the dry season; the deep, sticky clay discourages the passage of either vehicles or hoofed animals during the rains.

Given that the soils in the western corridor may be the most fertile in the ecosystem, the detour the migration takes on the way to the Masai Mara becomes understandable.[22] The interaction between rainfall and soil fertility can be seen in the distribution of both people and wildlife in the Serengeti ecosystem and its surroundings. The western limit of the migration coincides with a transition zone to lower-fertility soils derived from granites. The park boundary tracks the transition, leaving the less fertile agricultural and pastoral land between the park and the lake on the outside. In other words, the richer soils defined by the wildebeest forage preferences were incorporated into the park.

However, an area north of the Grumeti River endowed with high soil fertility and rainfall was left outside the park.[2] Nominally protected by the Game Division as the Grumeti and Ikoronogo Game Controlled Areas, where settlement and cultivation were permitted, their status was upgraded to Game Reserves in 1994 (see map 5.2). The settlements were forcibly removed, but the game reserves continued to be poachers' happy hunting grounds until leased in 2002 by the American financier and conservationist Paul Tudor Jones. As discussed in chapter 10, the Smithsonian research project on wildebeest reproductive physiology was carried out in the Grumeti Reserve in 2002–4. Under efficient management, the two reserves have been restored as outstanding extensions of the Serengeti ecosystem (more in chapter 12). Now better protected against poachers than Serengeti NP, the reserve's resident wildlife has rebounded, and wildebeest and zebra in the hundreds of thousands suspend their migration going and coming to harvest the abundant forage, as I have been privileged to observe for a month or two annually since 2007.

BURNING QUESTIONS

Unlike the short-grass plains, the Serengeti medium and long grasslands produce enough fuel to burn in the dry season. As a general rule, every grassland in Africa that can burn will be burned at least once a year.

Grazing and burning can significantly alter grassland composition and palatability. The continuing quest for green pastures contributes to the irregularity of the migration's dry-season movements, as already noted.[2]

In the Serengeti, if poachers or cattle rustlers returning from a raid don't start fires to cover their tracks, or careless motorists and campers don't set fires accidentally, the park authorities can be counted on to set them. It is part of an early burning policy intended to produce a mosaic of burned and unburned pasture while preventing more general and destructive conflagrations late in the dry season. However, these fires, once started, are simply allowed to run their course. Consequently, implementation often ends up burning most of the grasslands, including the preferred forage of the migrating wildebeest and zebra, as I observed in June–July 2010 and 2012.

When plentiful seasonal or postseasonal rains stimulate a widespread postburn flush, the northward migration can be delayed. Kenyans in the Masai Mara whose livelihoods are linked with the Serengeti migration often voice their suspicion that the Serengeti NP authorities deliberately try to keep the migration inside Tanzania. A more likely explanation is failure to implement the latest of a succession of park fire management programs.

ONWARD TO THE MASAI MARA GRASSLANDS

Knowledgeable ecologists consider the Masai Mara grasslands to be secondary, created and maintained by fire. Previously forested, Mara grasslands began to expand some four thousand years ago. That was around the time Cushitic pastoralists from the Horn of Africa arrived in East Africa.[23-25] From the Narok District in Kenya to the Simanjiro and Kiteto Districts in Tanzania, a pastoral lifestyle based on cattle herding has endured for at least the past two thousand years.[26,27] As the Masai arrived only within the past three hundred years, they represent simply the latest wave of Nilo-Hamitic pastoralist tribes.[28] Burning off the old grass to kill disease-carrying parasites (ticks), deny cover to predators, and create new grass for their livestock is a time-honored practice of pastoralist peoples. As noted earlier, lightning almost never creates wildfires in East Africa, so human intervention is the most likely scenario.

A similar sequence of events has recurred in the past century. According to hunters, traders, and explorers in the early 1900s, the Serengeti-Mara area looked much the way it looks today, with extensive open grassland and lightly wooded patches.[29] Then came the rinderpest pandemic that wiped out wild and domestic ruminants and people in the whole Serengeti ecosystem, followed by regeneration of dense woodland. But in the 1950s a rapid conversion to grasslands began, correlating with increasing wildebeest, buffalo, and elephant populations.

During the rainy decade of the 1970s, when the wildebeest population reached carrying capacity at 1.3 million, they consumed so much of the standing grass that fires became infrequent and trees were able to reinvade the Tanzanian grasslands. But in the Mara the opposite occurred: the area of grassland continued to increase.[30] Until the stream of wildebeest entering the Mara became a river in the mid-1970s, they only consumed a small part of the standing crop of dry grass. Moreover, unusually heavy rains in the 1950s and early 1960s produced so much long grass that the Masai burned it sometimes two and three times in succession. Hot fires pushed back the edges of the woodland and the *Croton* thickets. Elephants also played a major role in destroying woody vegetation. While over 80 percent of the Serengeti elephant population was poached in the seventies and eighties, the survivors found sanctuary in the Mara, where a high volume of tourists made poaching risky. Once masses of wildebeest began using the Mara grasslands from July to October or November, they left very little of the long grass elephants normally consume. The elephants had little choice but to browse woody growth. Following the 1988 CITES (Convention on International Trade in Endangered Species of Wild Fauna and Flora) ban on the ivory trade, they also increased rapidly, reinforced by immigration from surrounding areas of growing human populations.[29]

THE SHOW MUST GO ON

En route to the Mara grasslands, the migratory hordes have to cross the Mara River or its tributary, the Sand River. A cement bridge across the Sand River carries vehicles into and out of Kenya. Guarded by a police post, it has been restricted for tourists since Tanzania closed the border in 1977.

The migrants have no trouble crossing the Sand River, but crossing the Mara is another matter. The Mara continues to flow through the dry season, thanks to heavy rains that fall on the Mau Escarpment headwaters. Its banks are high where it flows across the deep black-cotton soils of the Mara Plain. Here even trivial-looking watercourses are cut deeply enough so that wildlife can cross readily only at passageways used year after year (fig. 6.5). Lined with low bush and trees, these

FIGURE 6.5. A mob of wildebeest descending a sloping bank of the Mara, about to cross.

watercourses divide the plains like hedgerows into giant paddocks of no more than a thousand or so hectares (see fig. 6.1).[31]

Most wildebeest cross the Mara at the fords created and maintained by the migration. But sometimes a column of wildebeest arrives between fords, perhaps at a spot screened from view by bushes, only to find—too late—a sheer drop of as much as 10–15 m confronting them. The mass coming behind forces the ones in front over the edge and so on, making a waterfall of wildebeest tumbling down the cliff. I have video footage of such an event filmed by my friend Paul Khurana that I show during talks on wildebeest. Miraculously, no one was killed or even seriously injured in the film sequence, despite tumbling head over heels, landing on one another, then having to jump over logs to enter the water (fig. 6.6). Apparently they all made it across and had no trouble continuing up a comparatively gentle slope to the grasslands beyond. However, the footage shows a crocodile capturing and drowning a zebra who was crossing with the wildebeest.

In August 2006, I filmed a crossing by several thousand wildebeest a little upstream of Governors' Camp. They came streaming down a sloping bank but, after swimming across, found themselves confronting a

FIGURE 6.6. This army came out of the riverine vegetation onto an embankment. Those in front were forced over the edge by the press of the hundreds coming behind. Photo by Paul Khurana.

10 m wall broken only by a few paths barely wide enough for one animal at a time to pass. Wet wildebeest struggling to climb the wall soon made it very slippery. Chaos ensued as wildebeest trying to recross ran into the oncoming horde of swimmers. The current was strong enough to sweep many tiring animals, especially calves, downstream. Presently there was an accumulation of dead and dying gnus. Perhaps 10 percent of the throng made it up the bank to the top. Most managed to retrace their steps.

That was just a drop in the bucket compared to what happened a month later. As many as ten thousand wildebeest drowned in five or six days during crossings not far downstream from the Mara Serena Lodge, according to Brian Heath, director of the Mara Conservancy. When I first heard of it, I suspected the crossings that day coincided with one of the floods that come barreling down the Mara following frequent thunderstorms over the headwaters. Not so. It was another accident similar to the one filmed by Khurana (the crossing illustrated in fig. 6.6). Although no witnesses have come forward, a made-for-TV film featuring this accident (*Wild Case Files,* Episode 5, Tigress Productions, UK) shows that the site of the crossings passed through dense scrub right to the riverbank, then down a narrow defile to the water facing a steep rocky slope on the other side. Once wildebeest had established this route, wave upon wave numbering in the thousands followed one another, with a thousand or more drowning each day. The next week bloated wildebeest corpses piled up against the bridge 20 km farther south, affording a supply of carrion that could stuff all the vultures in East Africa.

In the absence of video footage of the crossings, the production uses Khurana's video to create the virtual reality, and I am featured as "the guru of gnu" offering my interpretation of the event. My speculation that the blind-following instinct compelled each successive wave of wildebeest, guided by scent, to follow the same route to their deaths is the final comment before "Case Closed" flashes on the screen. So much for reality TV.

Huge Mara River crocodiles cluster near the traditional fords. Gorging on drowned animals caught up on boulders and downed trees is a leisurely business that often enough makes active predation unnecessary. The crocodile shown in figure 6.7 was so stuffed that he lay motionless, following a wildebeest cross-over only with his eyes.

The far side of the river was lined with vultures waiting to digest crops full of carrion so they could rejoin the scramble competition. Up- and downstream of major crossing points, their white guano was splashed on every rock and bare bank.

Floods, steep banks, and crocodiles are not the only obstacles the migrants have to face as they cross and recross the Mara in their quest for green pastures. Mass crossings are the main tourist attraction of the Masai Mara. Visitors who fill the lodges and hundreds of tented camps between July and the end of the dry season are very disappointed if they fail to see this famous sight. Beginning at first light, vehicles head for the river and settle down to wait (until breakfast time) for a crowd of wildebeest to arrive. The vehicles line up along both banks (like the vultures) end to end, often so close together that the animals have to funnel through the gaps (fig. 6.8). Absent any effective law enforcement, the driver-guides engage in scramble competition for choice sites. A good view of the action ensures fat tips. The most coveted reward is to see crocodiles, with just their eyes and nostrils showing, drifting out and pulling under a swimming gnu or zebra, or sometimes a little Thomson's gazelle.

FIGURE 6.7. A crocodile stuffed so full it merely watches passing wildebeest.

FIGURE 6.8. An obstacle course created by sight-seeing tour vehicles.

Fascination with predation and death is, I now believe, deeply engrained in the human psyche, and may well date to the time when our ancestors were on the menu of predators our size and bigger. Perhaps that explains why park visitors will sit and stare and stare and stare at lions lying fast asleep but will pass a herd of active antelopes with at best a pause to snap a few photos. Wildebeest and other prey species display what looks to me like a remarkably similar fascination with their predators. They stare at and even follow a walking lion while maintaining a safe flight distance. I think we and they still share the fascination with predators that preyed on our forebears.

REFERENCES

1. Hopcraft 2010.
2. Thirgood et al. 2004.
3. Rusch et al. 2005.
4. Wilmshurst, Fryxell, and Colucci 1999.
5. McNaughton 1990.
6. Olff, Ritchie, and Prins 2002.
7. McNaughton 1988.
8. Anderson and Talbot 1965.
9. Peters et al. 2008.
10. Anderson 1963.
11. De Wit 1978.
12. Vesey-FitzGerald 1972.
13. Maddock 1979.
14. Estes, Atwood, and Estes 2006.
15. McNaughton, Banyikwa, and McNaughton 1997.
16. Belsky 1984.
17. Trlica 2006.
18. Briske, Fuhlendorf, and Smeins 2006.
19. Herlocker 1976.
20. Breman and de Wit 1983.
21. McNaughton and Banyikwa 1995.
22. Olff and Hopcraft 2008.
23. Marean and Gifford-Gonzalez 1991.
24. Marean 1992b.
25. Marean 1992a.
26. Lamprey and Reid 2004.
27. Reid 2012.
28. Ehret 2002.
29. Dublin 1995.
30. Dublin et al. 1990.
31. Talbot and Talbot 1963b.

CHAPTER 7

Social Organization

Comparison of Migratory and Resident
Populations

While serving as resident naturalist at a Masai Mara resort some years ago, I was often asked to explain the erratic, "stupid" behavior of wildebeest at crossing points. They can behave rather mindlessly, I have to admit. Even with no obvious deterrent, like basking crocs (often ignored and nearly stepped on, actually) or a line of cars, a file or column will walk down to the water's edge, look ready to plunge in, only to turn suddenly and gallop back to the plain. When zebras are present, either leading or following a column of wildebeest, they are more likely to keep going.

My answer: the assembled wildebeest are a crowd of strangers. Except for mothers accompanied by dependent offspring, the leaders—all bulls as a rule—are anonymous. Arriving at the water's edge, those in front come to a halt. No one is prepared to take the plunge. "Why me?," each says in effect. "Why don't you go first?" Until some brave or foolhardy soul finally jumps in, there are likely to be repeated false starts. Zebras, on the other hand, are organized in cohesive family and bachelor units. When a herd comes to a crossing, the members stick together and provide mutual support. It's as simple as that.

How do I know the leading wildebeest are all bulls? By taking note of the leaders every chance I get. Whether filing over the plain, going to water, in traffic jams at river crossings and passageways through a belt of woodland, bulls are in the forefront (fig. 7.1). How come? Because males are the impetuous sex. They have no responsibilities toward other

FIGURE 7.1. A moving column of wildebeest with bulls in the vanguard, as usual.

wildebeest. They're testosterone-driven to compete with other males to find and breed females in estrus. During migratory movements, they largely suspend competitive activities (except during the mating peak; see chapter 10) but only until the next stop. As soon as an aggregation spreads out and begins to forage, bulls stake out temporary territories wherein they try to detain females. In short, bulls are the most impatient and controlling class of wildebeest. Being in the vanguard, especially while simultaneously migrating and rutting, may also be an advantage. Leaders intercept more passing cows than bulls farther down the line.

Females have been selected over the millennia to be careful because survival of their calves depends on maternal care and guidance. Poor mothers leave few or no descendants. So, naturally, cows are more timid and generally more alert to danger than bulls.

The inherent differences in risk-taking by male and female wildebeest remind me of gender differences in our species. Take women and men drivers, for instance. It is common knowledge (reflected in insurance rates for teens) that men are more aggressive drivers than women. Women, the primary caregivers, are naturally more cautious. As both species share the same androgens and estrogens, it is not far-fetched to suppose that testosterone affects male behavior in comparable ways.

Bull leadership of movements is not inevitable. At a river crossing, sometimes a mother looking for a lost calf or a lost calf looking for its mother will take the plunge and the rest will follow like sheep. Yes, just like sheep. Wildebeest are just as sociable as the domesticated sheep. It's instinctive, manifest at birth. The same "sociability genes" draw gregarious species to their own kind like iron filings to a magnet. Is there perhaps an actual, dedicated set of genes that predisposes species to be gregarious? I'm unaware that any such genes have been identified in any species genome, but quite conceivably herding species have inherited the same gene complex, that is, share the same genetic tool box.

BIRDS OF A FEATHER . . .

From what I've said about the members of a mob of wildebeest being strangers, except for mothers with their offspring, it may seem contradictory to be told that subgroups based on sex, age, and reproductive status are characteristic of wildebeest aggregations. This becomes apparent when you drive through a concentration while sexing and aging the animals within 50–100 m of the vehicle. Although individuals may not be personally acquainted, there is obviously a mutual attraction between animals of the same sex and age.

The first division is between males and females. From yearling class on, the two sexes are mostly separate. Whenever an aggregation is feeding or at rest, the separation is enforced by territorial bulls, who are intolerant of all other sexually mature males. Nevertheless, bachelor males can attach to sizable aggregations, and all classes migrate together.

Subgroupings are most distinct around the time of the annual calving season. Pregnant females associate until they have calved, then join nursery herds of mothers with new calves, while clusters of nonpregnant females are often found within aggregations of bachelor bulls. Beginning at nine months of age, weaned calves start to associate in a separate yearling class containing both sexes. However, some yearling females and a smaller numbers of yearling males can be found in the nursery herds, still tracking their mothers.

During their second year, yearlings divide into male and female subgroups. By the middle of their third year, it is hard to tell females from nonpregnant adult cows, and at least 80 percent of them breed during the rut beginning at twenty-eight months. Bulls, however, can still be distinguished from adults even as three-year-olds by their smoother, less knobby horns and slender build; these subadults also tend to form subgroups within aggregations of bulls. They mature at four or five years. Sometimes you even see groupings of old bulls.

ORDERLY MOVEMENTS

Why do wildebeest move in single file? This is another common question. Other sociable antelopes and zebra also travel this way. If you fly over the plain, you will see that it is crisscrossed by innumerable narrow paths. In single file, each follower lowers its risk of being ambushed, tripping, stepping into a hole, or encountering some other obstacle (fig. 7.2). Moving on a wide front, each animal confronts the unknown and is at greater risk. Following is made easier by the pungent scent of the glands between the wildebeest's front hooves. Followers can stay together on the darkest night guided by smell. It enables successive waves to take the same route. As I said about the drowning of ten thousand wildebeest reported in chapter 6, I suspect that tracking the scent trail of a preceding concentration may lead to repeating a disastrous river crossing.

When wildebeest in hundreds of thousands are migrating, they move in dense columns rather than single file. Wave follows wave over the same routes through belts of high vegetation, across watercourses and other difficult places, flattening broad highways through the grass. (See fig. 6.1.)

Grouping patterns are different on short and longer grasslands. Wildebeest grazing on short swards spread out, as though in one megaherd, with few clusters (absent intense territorial activity; see fig. 6.1). But when an aggregation enters grassland high enough to conceal a predator, the animals cluster together (fig. 7.3). Groupings of this kind are called feeding spreads.[1] I see this as the effort of the herd members to place other members between themselves and danger. I have argued elsewhere[2] that forming a group was prerequisite for ancestral solitary antelopes to leave cover and venture into the open. Feeding spreads may illustrate this motivation. At the same time, feeding in a dense aggregation is an efficient way to reduce high grass to a short sward.

You may ask, then, what is it that makes sheep, notoriously, follow one another? It must be instinctive, the product of natural selection, in wildebeest and all the other gregarious ungulates. I suggest that herding stems from the primordial fear of being in danger when you're all alone and away from any refuge. Put it this way: Wouldn't you feel safer in a group if you had to spend the night on the open plain with no safe refuge?

FIGURE 7.2. Typical single-file movement of Crater wildebeest returning from a grazing ground to the central plain.

FIGURE 7.3. When wildebeest enter high grass that could conceal a predator, they crowd together in a feeding spread.

RESTING FORMATIONS

Small nursery herds of wildebeest and many other antelopes lie in a so-called star formation (fig. 7.4).[3] It happens involuntarily as members settle next to one another, not facing but back to back. They end up facing in all different directions, which is useful for detecting any approaching danger. Wildebeest are not a contact species, like buffalo for instance. Only calves and their mothers lie touching; the rest keep at least the distance that can be reached by the horns. From a hilltop or airplane, bachelor herds can be distinguished from nursery herds by their wider, more regular individual spacing.

Aggregations of wildebeest choose the most open available landscape to spend the night. Groups numbering up to several hundred animals lie in linear bedding formations (fig. 7.5). The length of a formation may be as long as 50 yards, while the distance across spans no more than perhaps a dozen animals. Both sexes may lie in the same formation. The spacing between individuals is wide enough to allow animals to move in and out of the formation. As an antipredator strategy, the bedding formation is particularly adapted to the spotted hyena tactic of running at a bedding formation, then waiting to see if any potential prey lags behind. The wildebeest are able to disperse immediately without interfering with one another. (If they were to lie in a big circle, those in the middle would be unable to get away as quickly.)

RESIDENT POPULATIONS

The most obvious difference between migratory and resident populations is that in the latter females and young are distributed in small herds (sedentary-dispersed, while members of migratory populations move in large aggregations (mobile-aggregated).

Three resident populations of the western white-bearded wildebeest live in the Serengeti ecosystem: in the Masai Mara, the Western Corridor of Serengeti NP, and Ngorongoro Crater. A fourth subpopulation in the Loliondo District has been mentioned in some publications, but I am unaware of any recent firsthand observations. (Map 7.1.)

By definition, a resident population comprises a network of territorial males, each defending a particular property all or most of the time, often over a period of years. The network encompasses the area occupied by females, which live in semiexclusive small herds with home ranges that include a number of bulls' territories (fig. 7.6).[4,5] Noncompeting males are

FIGURE 7.4. A herd resting in typical "star formation," wherein the members face in all directions.

FIGURE 7.5. Time exposure of a moonlit linear bedding formation on the Crater floor. This configuration enables all members to disperse immediately in response to the hyena running-approach tactic.

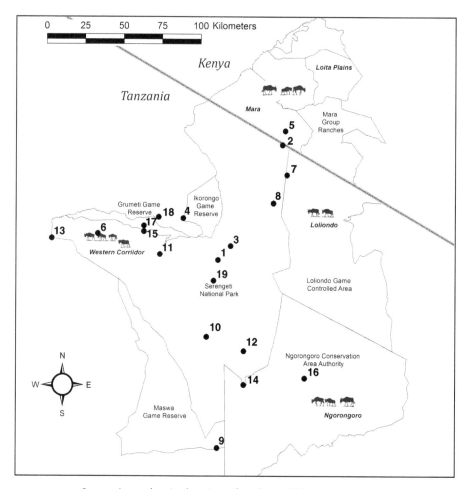

MAP 7.1. Serengeti map showing locations of resident wildebeest subpopulations, prepared by Grant Hopcraft. Also shown with numbered dots are Serengeti places: 1) Banagi; 2) Bolongonja park gate; 3) Serengeti Four Seasons; 4) Ft. Ikoma; 5) Keekorok; 6) Kirawira; 7) Klein's Camp; 8) Lobo Lodge; 9) Makao; 10) Moru Kopjes; 11) Musabi Plain; 12) Naabi Hill; 13) Ndabaka gate; 14) Ndutu; 15) Nyasirori; 16) Olduvai; 17) Sabora Plain; 18) Sasakwa; 19) Seronera.

kept on the periphery of the territorial network, usually on substandard and largely undefended range, where they associate in bachelor herds.

In migratory populations bulls defend temporary territories that they stake out whenever the movement of the aggregation they accompany comes to a standstill. Their herding activity creates pseudo-herds[1] of females and young out of the assembled mass while cutting out and marginalizing noncompeting males in the process (fig. 7.6).

FIGURE 7.6. Resident small herds on Ngorongoro central plain, each accompanied by a territorial bull.

As we shall see, the two forms are interchangeable. So whether these three populations should be classed as resident is debatable. However, they are definitely subpopulations with defined ranges different from the main Serengeti migration.

The Masai Mara Population

Certainly the main Mara population, which numbered >100,000 in the late 1970s[6], was migratory on a northeast-southwest axis, spending the rainy season on the semiarid short-grass Loita Plains and the dry season in the longer, wetter grasslands of the Masai Mara Reserve, a distance end to end of 135 km (90 mi.). An estimated 1,000–10,000 remained resident in the Masai Mara Reserve.[7] The population declined by 75 percent in the 1980s and 1990s, then stabilized to between about 12,500 and 20,000[8]. But by the mid-1990s this pattern appeared to have reversed, with resident wildebeest numbers stable at about 600 within the reserve but declining on the Loita Plains and group ranches.[9–11] During the past decade the Kenyan population has declined further, by perhaps 50 percent, suggesting that this population may now number no more than several thousand individuals. That includes several hundred that reside permanently within the Masai Mara Reserve (see summary in chapter 4).

Talbot and Talbot[12] wrote of a hilltop in the Masai Mara Reserve on the Egolok Plain, close to the Telek and Mara Rivers, where some 30 bulls were resident over the thirty-four months of their fieldwork between 1956 and 1962. At that time only 17,800 wildebeest were counted in the first aerial census of Narok District.[13] The senior game warden, Major Temple-Boreham, told them of one distinctively marked bull who had lived in this area for over thirteen years. The Talbots' description of lone bulls with permanent stamping grounds proves they were holding territories. Current estimates of the Mara population are in the low thousands; apparently resident small herds totaling a few hundred can be seen in the reserve and on neighboring group ranches in the months before the Serengeti migration arrives (author's observations, 2005, 2006, 2007). In other words, the migratory Mara population has been virtually eliminated.

The Western Corridor Population

Up to 20,000 strong, these wildebeest also engage in east-west seasonal movements. In the wet and early dry season, a territorial network and resident small herds frequent the *Balanites* tree savanna between Nyasarori and Kirawira. In the dry season, sizable aggregations graze the alluvial grasslands of the Ndabaka Plains at the western end of the park. The Serengeti population floods the Kirawira contingent in June and July. Some of the Serengeti wildebeest stay behind when the main migration heads northeast toward the Mara, but this only becomes evident the following calving season. The Serengeti recruits calve a month or two later than the "residents," who drop their calves in December/January.[14] (More under "Reproduction," chapter 10.)

The Ngorongoro Population

Like the Masai Mara and the "resident" wildebeest population of the Serengeti Western Corridor, the main Ngorongoro population is migratory. From 10 to 30 percent are truly resident, depending on rainfall amount and distribution from season to season and year to year. Although the Crater floor is a 250 sq. km (100 sq. mi.) grassland with just a few patches of woodland, resident subpopulations frequent areas with the best medium to short grassland. The main population moves about the Crater in the eternal quest for green pastures, on which they concentrate in thousands. Because the Crater is amazingly rich in the resources that sustain up to 20,000 large herbivores, it can support the large floating wildebeest population throughout the year. In important ways, Ngorongoro is a microcosm of the Serengeti ecosystem, affording a variety of similar grassland types on a greatly reduced scale. As noted in chapter 6, up to one-third of the Crater population actually emigrates over the western wall when rains fall in the Ol Balbal and Olduvai Plain without reaching the Crater. Here they occasionally meet and mingle with the Serengeti population.[15,16]

The diversity of Crater habitats reflects variations in rainfall, soil, and vegetation. A pronounced gradient occurs from the east to the west side of the Crater, a distance of 20 km (12 mi.). The outer side of Ngorongoro in the direct path of the northeast and southeast trade winds coming from the Indian Ocean receives 800–1,200 mm of rain, increasing with altitude, and is clothed in montane cloud forest. Inside the caldera, rainfall decreases with altitude and is greatly diminished at the Crater floor (at 1,850 m a.s.l. [6,000 ft.]), 460–615 m (1,500–2,000 ft.) below the rim. (Rainfall produced in rising vs. descending airmasses is discussed in chapter 2.) Precipitation averages 510–760 mm on the eastern and southern quadrant and only ~399–380 mm on the central and western quadrant.[17] Rainfall records at NCA headquarters on the southern rim average 630 mm.[16]

The fertile volcanic soil grades from deep and coarse to shallow and fine in the same downwind direction. The western wall is open grassland dotted with arid-adapted *Euphorbia nyikae* and *Acacia* species. The grasses range from tall to medium on the eastern and northern slopes to medium over the lowest slopes and adjacent floor to perennial short grassland on the alkaline soil around and west of Lake Makat.

DECREASE IN WILDEBEEST AND INCREASE IN BUFFALO POPULATION

When I was studying the Crater wildebeest in the early 1960s, five total aerial counts conducted between February 1964 and November 1966 recorded 10,500 to 15,000 wildebeest; 20,000 were counted in September 1964, the all-time record. The population remained at this level through 1980: the mean of 31 counts (mainly ground surveys) from 1964 to 1978 gave a mean population of 13,764 in the wet season and

FIGURE 7.7 (TABLE)   WET- AND DRY-SEASON GAME COUNTS OF CRATER
MAMMALS CONDUCTED BY NCAA STAFF, 2006–2012

Date	Wb	Ze	Bu	Gg	Ty	Hyenas	Lions	Lions True Number
2006 wet	3,385	3,484	3,106	764	1,549	47	13	72
2006 dry	4,688	3,786	2,045	421	793	46	49	73
2007 wet	2,673	3,043	2,623	201	726	29	19	59
2007 dry	10,730	4,778	1,741	557	1,577	119	12	53
2008 Apr.	8,325	5,433	3,689	613	582	95	14	58
2008 Sept.	10,526	5,485	3,133	482	770	102	23	50
2009 Apr.	12,006	5,099	3,303	505	1,151	125	23	57
2009 Sept.	8,921	7,181	2,116	548	945	85	10	57
2010 Apr.	8,505	5,626	3,007	552	661	82	10	52
2010 Sept.	10,235	4,816	2,452	650	1,242	85	10	50
2011 Apr.	5,111	3,722	7,952	353	532	128	34	62
2011 Sept.	6,689	3,958	5,372	326	790	135	15	56
2012 Apr.	8,008	4,942	3,609	316	1,193	0	10	52
2012 Sept.	8,901	3,255	2,340	306	1,119	195	8	58
Avg.	7,765	4,615	3,321	471	974	91	18	58
Max.	12,006	7,181	7,952	650	1,577	180	49	73
Min.	5,111	3,722	1,741	316	532	25	8	50
1999–2002	11,441	4,254	2,948	1,196	1,689			
2003–5	7,250	4,003	2,607	741	1,257			
Avg.	9,346	4,129	2,778	969	1,473			

(Wb = wildebeest; Ze = zebra; Bu = buffalo; Gg = Grant's gazelle; Ty = Thomson's gazelle)

SOURCE: Steve Mallya. Real number of lions provided by Serengeti lion researchers (Craig Packer, Ingela Jansson).

NOTE: Count averages 1999–2002 and 2003–5 indicate population changes (Estes, Atwood, and Estes 2006).

16,535 in the dry season.[15] No counts were made between 1980 and 1986. By the time counts were resumed, wildebeest numbers showed a steep decline. The mean from 1986 to 1998 was 7,573. Following a brief buildup to an average 11,441 wildebeest from 1999 to 2002, the population again declined to a mean of 7,250 between 2003 and 2005.[16] It has remained at that level through 2012: 7764, with a range of 5,100–12,000 through 2012 (fig. 7.7).

Thomson's and Grant's gazelles also declined from 1964 to 2005.[16] From ~3,500 tommy up to 1978, the average dropped to 1,400 between 1986 and 2005, a decline of 63 percent. Grant's gazelle declined from 1,559 to 960 between 1987 and 1992, a drop of 38 percent. After a brief increase from 1995 to 2000, their numbers fell to 740 between 2001 and 2005 (average of eight counts), down another 50 percent.[16] Average numbers continued to decline from 2006 to 2012, to 471, while the Thomson's gazelle mean populations fell to 974.

During this time span, counts of zebra have remained approximately stable (mean of 4,600, 2006–12), while the number of buffalo dramatically increased. Only a few dozen bachelor bulls resided in the Crater before 1970; the big, mixed herds stayed in the glades and thickets of the surrounding montane highlands. But over 1,000 were counted inside Ngorongoro as early as 1973. Still, the mean number of buffalo in the Crater up to 1980 was only 523. Then from 1986 to 1992, the mean increased fourfold, to 2,392, and rose another 16 percent from 1993 to 1998. From 1999 to 2005, though counts ranged as high as 5,000 and as low as 1,000, the average has remained at 2,800 ± 590. The 2006–12 counts record a 15 percent increase, to an average of 3,300 (range 1,741–7,952) (fig. 7.7).

Despite these major changes in species populations, Ngorongoro's mean total herbivore biomass had not changed significantly: 3,294,450 kg ± 193,615 kg, amounting to 10,982 kg per sq. km (based on twenty-seven counts).[18] Because buffalo weigh three times as much as wildebeest, they had replaced the wildebeest as keystone species despite a much smaller population.

What had brought about conditions favoring the buffalo and at what cost to the wildebeest and gazelle populations? A combination of factors has resulted in a series of changes:

1. The NCA's policy of excluding fire, pursued from the 1960s to 2005, promoted a shift from predominately medium and short grassland to long grassland. The shift favored buffalo, which can digest tougher grasses than wildebeest or Thomson's gazelle. Yet one experimental burn of 10 sq. km late in the 1962 dry season of floor and slope grassland overgrown with *Leucas* bushes and *Gutenbergia* weeds attracted one-third of the wildebeest population to the lush postburn flush in November and over three-fourths of the population the following February. Wildfires that burned much of the northern and eastern slope grasslands in 1963 and 1964 were heavily used by wildebeest, zebra, and gazelles during the following rainy seasons. This was the reverse of the usual seasonal rotation: the hill grasslands, dominated by *Themeda triandra* and *Andropogon greenwayi*, which stayed

green longer than the floor grasslands, were normally grazed early in the dry season and mostly unoccupied during the rains. What the wildfires showed was that burning the long grasslands could increase the pastures regularly occupied by wildebeest from 104 sq. km, to some 233 sq. km, thereby including almost the Crater's entire grassland area.[15] A regular burning program finally began in 2007.[19]

2. Increasing diversion of water flowing into the Crater from surrounding highlands for construction projects (new lodges) on the Crater rim, with an attendant increase in the human population, including tripling of the Masai to over 60,000 residents, led to earlier drying of the grasslands. Reduced rainfall did not account for this, as early drying occurred from 1994 to 1997 despite rainfall above the normal average of 630 mm at NCA headquarters. It was followed by the near-record El Niño rains of 1998 totaling 1,614 mm. Subsequent major road works on the Crater floor further altered natural drainage patterns, impeding overflow from the swamps onto adjacent grasslands.[16]

3. Major changes in the composition and growth form of Crater grasslands have occurred since a vegetation survey published in 1972,[17] coinciding with the declining wildebeest population. Areas in the southern and western plains that had been short and medium grassland three decades earlier were replaced by tall growth that was only lightly utilized (Herlocker pers. comm.). These areas of short alkaline grassland and short to medium *Cynodon* grassland had been major wildebeest concentration areas during the rains, comparable to concentrations on the Serengeti short-grass plains.[15]

   Surveys of the same sites sampled by Herlocker in 1972 showed that fewer species now dominate the Crater's grasslands. Two species that once dominated large areas, *Sporobolus homblei* and *Digitaria macroblephara,* the latter one of the most abundant grasses in the 1960s, have disappeared.[19] Meanwhile, *Cynodon dactylon, Andropogon greenwayi, Sporobolus ioclados,* and *Chloris gayana* have increased, especially in areas formerly dominated by *D. macroblephara.* The first two are abundant in pastures with moderately heavy grazing, whereas *Chloris gayana* pastures are underutilized except after burns. *Digitaria abyssinica,* once a major dominant on the Crater floor, has decreased to

FIGURE 7.8. Crater floor overgrown with *Gutenbergia cordifolia* in 2002, a weed that invades and shades out grassland.

varying degrees but remains abundant in recently burned grassland subject to heavy grazing.

4. Repeated infestations of the weeds *Gutenbergia (Erlangia) cordifolia* and *Bidens schimperi* have reduced the available forage for herbivores in large areas (fig. 7.8).[16,20]

Though the connection between the changes in the Crater's grasslands and reductions in the wildebeest population have not been studied, the results of research on Serengeti grasslands, discussed in chapter 6, strongly suggest that the changes came about because the wildebeest's impact on its food supply, which promoted the most palatable grasses best adapted to heavy grazing ("self-facilitation"), was drastically reduced.[21-23]

OTHER CHANGES

Change to taller grass improved habitat preferred by black-backed jackals at the expense of golden jackals, which were codominant when short grass covered much of the Crater floor. The wide area covered by *Chloris gayana* is heavily infested with ticks, for which an important host is the buffalo. The ticks are carriers of babesiosis, a disease that killed five of the Crater's lions and five rhinos following a severe drought in 2000. Fires routinely set by the Masai elsewhere keep their cattle largely free of ticks while also improving grazing quality.[24] Some 1,500 buffalo died but of starvation, not disease. Counts in 1999 and 2000 indicated that the population had increased to 5,000, the highest number ever recorded in the Crater. A condition check early in 2000 found the buffalo was in much poorer condition than the antelopes, and by October virtually all the grazing on the Crater floor had been consumed.[16] It seems likely that the number of buffalo exceeded the Crater's carrying capacity, making them more vulnerable to drought than the other herbivores.

Repopulation of the Crater in the 1970s by warthogs *(Phacochoerus africanus),* which had been unrecorded for many years, was undoubtedly helped by the shift to tall grass. When I moved into the Crater in 1963, no one could remember seeing any warthogs. Yet when my campsite on the Munge stream was being leveled, we found a warthog tusk. This suggested warthogs had formerly lived on the Crater floor.

The reduced wildebeest population was followed by a reduced Crater hyena population. Estimated at 385 adults in the 1960s, the highest density anywhere, their number was down to 139 adults in 2000.[25,26] They increased again, to 228, in the late 1990s, tracking the wildebeest population increase during that time. As their main prey were adult wildebeest and buffalo calves, these results indicated to the hyena researchers that the hyena population was controlled by the prey population, not the other way around. The 2006–12 average count was 104 (range 29–180).

REFERENCES

1. Watson 1967.
2. Estes 1974.
3. Walther 1966.
4. Estes 1966.
5. Estes 1969.
6. Sinclair and Norton-Griffiths 1979.
7. Estes and East 2009.
8. Ottichilo, de Leeuw, and Prins 2001.
9. Homewood et al. 2001.
10. Sinclair and Arcese 1995.

11. Sinclair 1995a.
12. Talbot and Talbot 1963b.
13. Stewart and Zaphiro 1963.
14. Ndibalema 2009.
15. Estes and Small 1981.
16. Estes, Atwood, and Estes 2006.
17. Herlocker and Dirschl 1972.
18. Runyoro et al. 1995.
19. Amiyo 2006.
20. Henderson 2002.
21. McNaughton 1985.
22. McNaughton 1986.
23. Owen-Smith 1988.
24. Trollope and Trollope 2001.
25. Höner et al. 2002.
26. Höner et al. 2005.

CHAPTER 8

Male and Female Life Histories

The natal sex ratio of wildebeest is equal.[1] It remains equal as long as both sexes stay with their mothers and/or in herds of females and young. Calves remain dependent on their dams until weaned by nine months. From then on through the next calving season (February–March), most calves separate from their mothers and associate as a class within aggregations. But some, mostly females, continue to follow their mothers as yearlings through the next calving season.

Despite minimal sexual dimorphism (fig. 8.1), territorial bulls can readily tell which is which (by scent if not by sight) and routinely evict males from herds of females and young. In resident subpopulations, being chased from the maternal herd and familiar home range, often repeatedly, is clearly stressful. It is probably less so in mobile aggregations, wherein there are no stable female herds or fixed home ranges. In either case, the males end up in bachelor herds. Higher male mortality rates begin with sexual segregation, leading to adult sex ratios favoring females. In resident populations the ratio favors females 3:2 or 2:1. The more dissected the habitat (mixed grassland and woodland), the more skewed the adult sex ratio. In large migratory populations, though, adult sex ratios have been considered nearly equal.[1,2]

As explained earlier, females, mothers in particular, are naturally more alert to danger than are males. As long as the sexes are associated, males benefit from female vigilance. Separate bachelor herds are marginalized by territorial males, often in substandard habitat where they

FIGURE 8.1. Similar appearance of adult male (l) and female wildebeest; note female penile tuft.

are more vulnerable to predators. Lower male vigilance is a contributing factor. However, minimal sexual dimorphism enables bachelors of all ages to join large aggregations, thereby gaining access to preferred grazing. Enabling access to green pastures for their sons may have been one of the advantages of reduced sexual dimorphism that led females to look more like males (refer to chapters 1, 3, and 4).

Though no bigger as newborns, male growth rates exceed female rates from then on—and males keep growing another two years after females reach adult weight.

The wildebeest growth curves shown in figure 8.2 represent constants (the von Bertalanffy growth model) that apply not only to wildebeest, but to all the other ruminants featured in this book. Nick Georgiadis applied the model to eleven African ruminants, ranging in mature size from 20 to 600 kg, in his paper "Growth Patterns, Sexual Dimorphism and Reproduction in African Ruminants."[3] The boxed text based on the article abstract, summarizes the main results.

The wildebeest's 20 percent size difference is not obvious. It is similarities in coloration and adornments that enable mixed aggregations. Take the horns for instance. Male horns are more robust—in every bovid—as a consequence of peer competition, but mimicking horn shape is what counts. Female horns have not been subject to the same rigorous sexual selection. Yet I see and take note of females with broken horns while sexing and aging the passing parade but rarely see more than one or a few in a thousand. The rate is higher among males, even though their horns are a lot harder to break. The proof: I started research in

Ngorongoro by collecting wildebeest skulls, figuring this would be an easy way to determine the adult sex ratio. But except at fresh kills, I only found the remains of male skulls, consisting of the horn boss and attached skull remnant. It eventually dawned on me that hyenas could crunch the horns along with the rest of females' skulls.

LIFE EXPECTANCY

Based on postmortems of shot samples and wildebeest that died of natural causes, Watson[4] estimated male and female life expectancy of the Serengeti migratory population at eight and seven years, respectively, with maximum longevity of sixteen years. The greater demands of maternity explain the shorter average in females. With first calves produced at three years, cows would have at most five or six offspring. Reproductively successful bulls, beginning at the age of four or five, could of course sire dozens more offspring. Accordingly, bearing the greater nutritional demands of suckling males would have a higher potential payoff. In some ungulates, more males are conceived when females are in peak condition and more females when they are in poorer condition. Such an arrangement would help a population to recover quickly from a disastrous decline. But I know of no evidence that this applies to wildebeest. By avoiding the energetic costs and dangers of migration, resident wildebeest tend to grow bigger and may have a life expectancy one to three years longer.[5]

MALE LIFE HISTORY

Life in a Bachelor Herd

All ages, from yearlings to old bulls past their prime can be found in bachelor herds. Wildebeest in all-male herds seem inactive compared to most associated antelopes. Bachelor topi, hartebeest, Grant's and Thomson's gazelles, and impala have frequent tournaments as peers test one another in play and real fights. This difference seems odd, considering that territorial wildebeest are among the most interactive and aggressive antelopes. Yet spacing between landholders is considerably closer compared to associated species, arguably reflecting greater tolerance of their territorial neighbors.

The members of a wildebeest bachelor herd are amiable but not as companionable as females. There is no apparent rank hierarchy or other

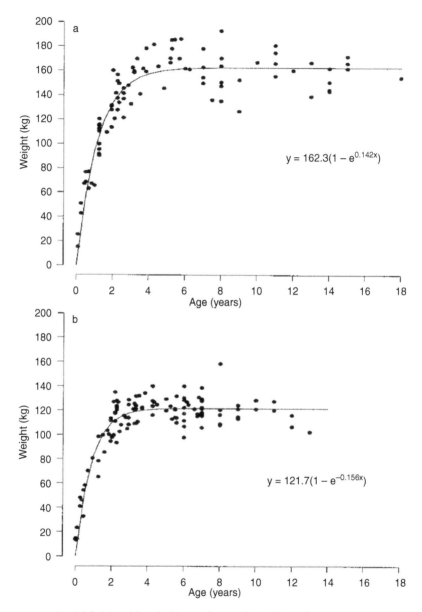

FIGURE 8.2. Male (a) and female (b) growth rates. From Hopcraft 2010.

(i) Growth rates are systematically related to adult body size. Larger species grow relatively slower but absolutely faster than smaller species. (ii) Male and female growth patterns differ in several ways. Males of sexually dimorphic species attain a mature weight that is greater than that of females by growing slightly faster and, more important, by continuing to grow after female growth has stopped. (iii) Female growth is not affected by the degree of sexual dimorphism, only by adult weight. Females of all species first conceive at about 80 percent of adult weight. The maximum potential number of offspring per lifetime (about 14) for females is also independent of adult size. (iv) Male growth is affected by the degree of sexual dimorphism (SD), suggesting that sexual selection has acted on males rather than females. Males of markedly dimorphic species take longer to reach mature size than males of monomorphic species having the same mature size as females. Thus growth is relatively slower among males, depending on the degree of SD. (v) Adult sex ratio is negatively related to SD. This suggests that males of dimorphic polygynous species generally suffer higher mortality rates than males of monomorphic monogamous species. Growth rate, metabolic rate, and longevity may be linked to survivorship costs by the degree of SD: the greater the sexual dimorphism, the higher the costs.

organization. There is little or no serious fighting and little play. The gnu bachelor herd is thus a comparatively stodgy club. Reproductive physiology of gnu bulls, based on sizable shot samples of Serengeti animals, may help explain bachelors' mild manners.[1] Spermatogenesis begins in two-year-olds, when bulls attain a weight of about 160 kg and a testis mass of 20 g. Mean adult testis mass of about 130 g is attained at the end of the third year or early in the fourth. But bulls five years and older have a testis mass of 153 g. If that's what it takes to enable bulls to compete successfully for territories, then a good year must pass after bulls reach adult size before they have the cojones to compete for territories.

Nevertheless, most if not all of the fixed action patterns associated with territorial behavior are performed by immature males, although typically at low intensity. The actions associated with marking the territory, detailed later, are "practiced" by bachelors before they have any property of their own. Young bulls perform the main steps seen in the challenge ritual between neighboring territorial males, including horn contact and cavorting. Sexual behavior can be seen in bachelor herds in the rutting season, or whenever females in estrus are present (see fig.

10.12). Finally, the advertising call of territorial males is attempted even by yearlings, in the "falsetto" of a calf, while the bass tones of two-year-olds go high and "crack" like the voices of adolescent boys. The resonant basso grunt is finally achieved at about two and a half years.

As bachelors advance through the age classes, horn development keeps pace; the gap between them closes between two and three years, and the bosses, essential for withstanding hard knocks, become prominent in the fourth year (fig. 8.1; see also fig. 4.3). Age can be estimated by the depth of the gap between the horns, which is a measure of boss development. Body size and horns reach their maximum development when bulls mature at around five years. Individual differences in horn development and physique (up to 12 percent of body weight) are evident. From then on wear and tear take their toll. Horn tips wear down, the gap between horn bosses widens again (fig. 8.3), and bulk lessens with age.

Although segregation from females relegates bachelor herds to substandard habitat most of the time, even in sedentary populations bachelors have the opportunity during the dry season to join temporary aggregations of resident herds on the shortest, greenest grass. A condition check of Crater wildebeest made at the end of the 1963 dry season indicated that bachelor males were in as good if not better condition than females with young (43 vs. 35 percent). In contrast, only 16 percent of the sampled territorial bulls were judged to be in good condition.[6]

Bachelor herds, often in association with zebras, also have the advantage of entering tall and bushy vegetation avoided by females. Risky business, but *qui a peur du méchant lion?*

The ratio of bachelor to territorial males in the Ngorongoro population changes with the seasons. In the 1960s there were something like 3,000–3,500 males over two years of age in a population averaging ~12,500. An estimated 1,000–2,000 bulls held territories outside of the rut and late dry season, the times of maximum and minimum territorial activity. This amounted to 8–16 percent of the total population, and perhaps one-third to two-thirds of the adult males. For comparison, Talbot and Talbot[7] estimated that single (= territorial) males made up over 8 percent of the Serengeti population but only 16 percent of the adult males. His figures were calculated on the basis of a nearly equal adult sex ratio.

Although exact percentages for Ngorongoro bachelor and territorial wildebeest cannot be offered—and would apply only at one time and place anyway—it is safe to assume that up to 50 percent of the adult males may be relegated to bachelor herds at any time, not because they are unfit, but simply because they are supernumerary. Presumably the ratio depends largely on the number of adult males in relation to the area of suitable habitat; in short, the ratio of haves to have-nots is also density-dependent. Bachelor herds may be viewed as reserve banks that are drawn upon to fill vacancies in the territorial network. At the same time, bachelors are the main source of competition to established males for continuing possession of their territories. This reserve of fit bulls promotes the phenomenon of permanently occupied territories even when mating opportunities are few to none.

The Individual Territory

As there are no visible boundaries to show where one territory ends and the next one begins, it is far easier to measure the spacing between bulls than it is to delineate a single territory. In fact, the only concrete evidence of an occupied territory, apart from the owner's presence, is a bare spot the size of a tabletop, liberally covered with dung—the bull's so-called stamping ground. Only if some natural or artificial barrier—a stream, tall vegetation, a roadway or fence, for instance—intervenes does the boundary become definite and fixed. Yet even though it appears entirely indistinguishable from the surrounding featureless plain, a territory is a definite piece of real estate, every inch of which is known to the owner. But the only way for an observer to determine the limits is to keep a record of the positions occupied by a known individual relative to a fixed reference point. After first trying numbered stakes, which zebras rubbed against and knocked over and hyenas chewed up, I put out numbered whitewashed stones, usually just on the edge of the stamp.

Territory owners were so rarely found more than 100 m from their markers—assuming they were found at all—that bulls any farther away were counted as out of their territory. In 1,291 observations, 21 known bulls were found within less than 100 m of their stamping grounds on all but 146 occasions (11.3 percent), and this includes the times when they were temporarily absent altogether. In a sample of 217 observations of 12 bulls where distance from the marker was visually estimated and plotted, the frequencies at different distances were as follows: 0–30 m, 37.8 percent; 30–50 m, 34.6 percent; 50–75 m, 20.7 percent; >75 m, 6.9 percent. These bulls spent more time on the stamp—18 percent of the observations—than in any other single place From these records it became clear that the stamping ground is the nucleus or heart of the territory—though not necessarily the geographic center.

Since no actual boundary, visible or invisible, could be said to exist between territories, it could only be surmised that the border lay somewhere approximately halfway between two adjacent bulls, as a vague zone or no-man's land. Except for the outermost bulls in a network, each individual was surrounded by neighbors and had common borders with at least three and up to five or more. Under these circumstances, territories could not be circular, as often conceived; if anything, they would be polyhedrons like the cells in a honeycomb. A territorial network may be thought of as a mosaic made up of pieces of varying shapes and sizes, each piece being an individual territory.

The Crater's central plain was usually frequented by separate small nursery herds. Here the spacing between bulls at any given instant was anywhere from 1 m (when neighbors were interacting) to over 300 m; individuals were of course constantly changing position relative to one another. That made it hard to estimate the average spacing even where there were no herds to complicate matters. My impression was that the spacing between males in prime habitat was about 60 m. But actual measurements indicated that this was nearer the minimum average than the general average.

To get a better idea of the minimum spacing and population density of territorial males, I made three road counts along the same 19 km (12 mi.) circuit of the central plain, including bulls within 150 m of either side of the vehicle (detailed in Estes).[6] A month before the rut (April 30, 1965), 249 territorial bulls were counted, giving a density of 123 per sq. km (319/sq. mi.), with an average spacing of 155 m (169 yd.). During the rut, two samples taken on May 30 and June 6 had a mean density of between 57 and 85 territorial bulls per sq. km (147 and 221/sq. mi.), spaced 120–140 m apart on average. The density in this area at other times of the year, except in the late dry season, appeared comparable to the April figures.

When compared with most other antelopes of open habitat, the spacing between territorial wildebeest is seen to be much closer. For instance, spacing between territorial Coke's hartebeest in Nairobi National Park, the densest known population, averaged 500 m (minimum 200 m), equal to a density of only 2.7 males per sq. km. The two gazelles in Ngorongoro, though much smaller than wildebeest, defend much bigger territories. Thomson's gazelles maintain a spacing of roughly 250–350 m, while territorial Grant's gazelles keep 0.4–0.8 km (1/4–1/2 mi.) apart.[8]

FIGURE 8.3. Old bull branded E stands calling on the territory he held in 1964 and 1965.

On Becoming Territorial

During the rut in Ngorongoro Crater, I have seen males as young as twenty-eight months attempting to defend a territory. As a subadult couldn't stand up to challenges by mature neighbors, the youngster would move away, only to return when the threat receded. Similar tactics are deployed by old bulls rejuvenated by the rut's frenzied activity and by bulls rendered unfit by broken horns, but those with both bosses intact can still bash heads with their neighbors (see old bull E [for "Eggling"] in fig. 8.3). Some four-year-olds participate but apparently lack the stamina to hold territories day after day in the thick of things. In my samples of territorial males, the great majority are in their prime but middle-aged, more than five and probably under ten years.

In a resident population, how does a newcomer win a place in an established territorial network? The members of this very exclusive club occupy all the best wildebeest habitat. I think of it as a form of Hamiltonian (vs. Jeffersonian) democracy. Only property owners have the right to vote, that is, participate in the mating sweepstakes. So suppose you are a vigorous five-year-old bull searching for a place of your own, how do you go about it? Could you just confront a territorial individual and oust him in a knock-down, drag-out fight? It would take an exceptionally aggressive and determined bachelor to beat up a fit bull who has occupied a piece of land for months if not years. His claim has been confirmed every day by scent-marking and visual displays and in chal-

lenge rituals with each of his neighbors, all of which reaffirm his property rights. The psychological advantage of ownership ("my home is my
castle") gives the owner a strong home court advantage. Maintaining
this advantage favors bulls that stay on territories even during months
when there are few to no mating opportunities.

Thus the homesteading bull faces a dilemma: in order to hold his
own in the company of established bulls, he must first become master of
a territory. But he can only join the club when he can stand up to other
bulls. The sexual-territorial drive is tightly linked to property. So any
bull without property, and even a territorial male away from his place,
lacks the necessary aggressiveness to challenge bulls on territory. In
effect, a bull without property is a psychological castrate.[6]

The preferred strategy is to look for a lightly defended spot and try
to stake it out. The easiest way is to set up on the periphery of the network, but the reason that's easy is because females rarely stay there.
There are also places within the network that are marginal because of
proximity to cover that could conceal a predator, as along streamside
vegetation or near a stand of tall weeds. A prospecting bull can find
such a spot by hanging around and assessing the push-back of the nearest territorial bulls. If chased, he will run away, but if it's not done very
aggressively or if he's strongly motivated, he will keep coming back.
Persistence pays off. The aggressive response of bordering property
owners is gradually fatigued—that is, the residents gradually become
habituated to the interloper's presence. They chase him less persistently
and/or less often.

As the aggressive response wanes, so the self-confidence of the newcomer waxes, along with a growing proprietary interest in the place he
frequents. Soon he dares to call, to move about actively in the rocking
canter, and to herd approaching willdebeest and even bully transient
males. At this point he is acting like an established territorial bull in
all but one respect: he will not stand up to a neighbor's challenge. If a
bull walks toward him, calling and nodding his head, he will move
away, only to circle back to his place the moment the other turns
around. If nursery herds are in the vicinity, an indecisive bull will
often apparently feign preoccupation with herding and chasing them.
Seemingly oblivious of his approaching abutter, he will canter off
toward them, head high and calling, but if actually pursued, he will
run away.

When the day finally comes that the newcomer feels completely
secure in his right of occupancy, he is by way of achieving full member-

ship in the territorial establishment. Equality is reached when the repelling force of his aggressiveness, which has grown in direct proportion to increasing self-assurance, becomes equal to the force exerted by the bulls on the surrounding territories. At first his force may be equal only within a short distance of his stamping ground, but with time his domain expands until it reaches the maximum for this particular individual in his particular circumstances.

How long it takes a newcomer to become established also depends on such factors as terrain, vegetation, season, and how closely resident bulls are spaced; very important, on the temperament and motivation of the individual and of his prospective neighbors; and no doubt also on age and previous experience. Given a favorable combination of circumstances, a bull may achieve the requisite self-assurance within a few days or even hours. Less favorable conditions may cause the process to drag on for weeks, and an unfavorable combination can prevent the process all together.

Territorial Tenure and Attachment to a Particular Spot in a Sample of Known Bulls

The evidence gathered from marked bulls in Ngorongoro Crater demonstrates a continuing attachment to the same piece of ground while revealing at the same time great variability in length of tenure.[6] The sample included a total of 69 males, of which 13 were identified by natural peculiarities, 23 by brands and ear tags, and the rest by photographs of their stripe patterns. Twenty-six were lost from the sample within a month or less, as follows.

Among 6 of the branded animals, 4 disappeared within less than two weeks of being marked; only one was ever resighted. The other 3 were presumably taken by predators. Two were subadults, marked as yearlings, that were found on territories for only a day or two during the May 1964 rut, when they were no more than twenty-eight to twenty-nine months old. Both were subsequently resighted in bachelor herds. (A third such male became territorial again several months after the rut and was kept under observation thereafter until the end of the study.) Three of the naturally marked animals were identified for the first and last time during the 1964 rut, and all abandoned their territories within a week or less. Of the males whose stripe patterns were photographed, 17 were never resighted, either because they were on temporary territories (7 were photographed during the peak rut), or had not been on their own territories when the photographs were taken and their locations were plotted, or their stripe patterns were so undistinguished that they could not be told apart from other animals with any certainty.

This left a sample of 43 bulls on "permanent" territories. For the purpose of the study, a male that remained for at least a month on the same territory, or repeatedly returned to the same place after leaving it, was considered to belong to the population of "permanently" established territorial males. The sample therefore is not of all adult Ngorongoro males but only of males belonging to the permanently established territorial network typical of a nonmigratory population.

All of the males in the sample, except the third-year male already noted, were mature when first observed and had occupied their territories for an unknown length of time prior to their inclusion in the sample. The results therefore suggest only the minimum period of occupation. The bulls performed as follows.

One-third (14) were no longer on their territories at the end of the study. These were branded animals, of which 5 were subsequently resighted and known to have survived. The disappearance of the other 3, which were not seen again, was probably due to predation. The other 6 belonged to the photographic sample and could be identified only as long as they stayed put; there is no clue to their fate. Of these 14 animals, 9 left their territories in May or June, during or soon after the peak rut. Four left their territories within six months of the time first observed (one within the first month).

None of the 29 males still on territory at the end of the study had been in residence for less than ten months. Of 26 males whose histories were known for at least one year, 20 (77 percent) were still in place at the end of the study. Of 16 males whose histories were known for eighteen months or more, 11 (67 percent) continued to hold the same territories.

Of 13 males that were in the sample for two years or more, 7 (54 percent) were still on the same territory at the end of the study, 2 were on new territories, and 3 were in bachelor herds. The thirteenth animal, a male with both horns broken off at the curve, which was classed as old when darted and branded "E" (for Eggling) in June 1962 (see fig. 8.3) remained for most of the next two years in a bachelor herd. Then, at the onset of the rut in May 1964, this bull took up a territory and was still on it at the end of June 1965.

Attachment to Place

The fact that a male still had the same territory, however, did not necessarily mean that he had remained continuously on it. One bull was absent for some six months during the dry seasons of 1963 and 1964 and presumably spent those months in a bachelor herd. Two others were sometimes missing for periods of several days up to a month. However, long absences were the exception to the rule.

The mere presence of a bull in a particular place shows that it is an occupied territory. Absence, conversely, invites encroachment by the neighbors or by an outsider. The need to be present may account for the fact that the known individuals seldom left their territories for more than a few hours at a time. In 1,291 visits to the territories of 21 different bulls at all seasons, they were recorded as present 89 percent of the time. This shows the importance of staying put. Former, displaced occupants of the site pose a particular risk. One bull who was absent from his territory overnight after being immobilized and marked found it occupied on returning the next day and only regained it after a hard fight.

The first bull I immobilized and branded lost his territory within two weeks of the incident. Thereafter, he was seen in a large bachelor herd two miles away. Although his former place was promptly claimed by his neighbors, I saw him in the immediate vicinity nineteen times over the next six months. Twice he was seen fighting and once in a challenge ritual. During the following year he occupied land close to his old territory three times—only to lose it in a matter of days. On a visit to Ngorongoro after an absence of four years, I was gratified to see that he had regained his old place.

The former neighbors of a bull seeking to reclaim his lost territory seem determined to prevent it.[6] Here's another example that demonstrates attachment to a specific piece of real estate, well named *Ortstreue* (fidelity to place) in German. A marked bull whose territory was taken over during an unusually prolonged recovery from darting kept turning up for the next two years as close as possible to his former property (marked with a numbered white stone). The ferocity with which his former neighbors chased him away suggested a gentleman's agreement to keep him out. But then a newcomer would probably lack the motivation to try his luck in the middle of an established network, so maybe it wasn't familiarity that bred contempt.

Mere presence may be enhanced, still in a passive way, by standing as opposed to lying. In Ngorongoro, territorial bulls very often stand motion-

FIGURE 8.4. Static-optic advertising. A bull stands and ruminates on his stamp in a distinctive head-high pose that may be maintained for an hour or more at a time. A territorial Thomson's gazelle shares his stamp.

less for long periods, especially in midmorning and early afternoon, as though asleep on their feet but with heads raised (the head-high posture). But they apparently do not sleep, although they close their eyes and may doze for a few minutes at a time. Usually they stand chewing their cud.

If there is a breeze, most bulls face into it. Standing in a conspicuous position like this is a widespread form of territorial advertising called static-optic marking (fig. 8.4).[9] It applies even more obviously to the hartebeest's and topi's well-known habit of standing motionless on termite mounds or other elevations (see fig. 3.1). Since this position is also clearly advantageous for surveillance, the functions of seeing and being seen can be simultaneously served. Other, more dynamic territorial displays include depositing dung on the stamping ground preceded by vigorous pawing, kneeling and horning the ground or vegetation, lying down and rolling, calling, approaching trespassers with head nodding and in the rocking canter, followed by herding or chasing (fig. 8.5).

Relations between Territorial Neighbors: The Challenge Ritual

I pointed out earlier that territorial advertising has an intimidating or challenging effect on other males according to their own status.

FIGURE 8.5. Territorial bull going to meet approaching wildebeest in the rocking canter, calling at the same time.

Bachelor males, and even territorial males away from their own property, would almost invariably move meekly away when approached aggressively—though mature individuals might stand up to bulls that crowded them too impetuously.

A bull's own neighbors were by no means so readily intimidated. In fact, they often reacted to demonstration marking or threat displays as if their own territorial integrity were being challenged and they were bound to react. They might respond in kind, or they might actually invade the neighbor's territory to demonstrate face-to-face, leading sometimes to a skirmish. These encounters also took place without prior demonstration and in fact proved to be a basic feature of gnu territorial behavior. Bulls in the sample of known individuals were found interacting with their neighbors an average 6 percent of the times they were observed, spending a minimum of forty-five minutes a day in this pursuit. Engagements on moonlit and even on dark nights were also observed. It appeared that every bull had at least one encounter, and often several, with each of his neighbors in the course of a day.

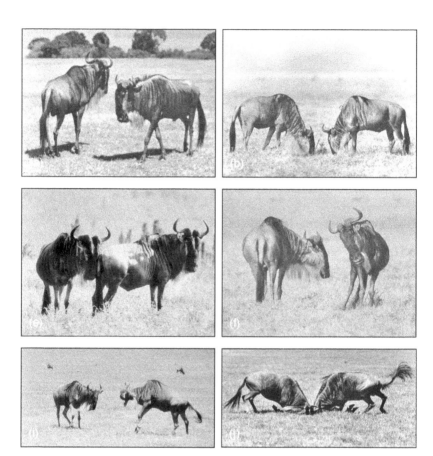

FIGURE 8.6. *(Two-Page Spread)* Steps in the daily challenge ritual between territorial neighbors: a) Approach to bull standing in side-on presentation, head raised. b) Both engage in displacement grazing, a neutral, unaggressive behavior. c) One paws the ground, preparatory to defecating, which is inconspicuous in the gnu. d) Bull in foreground is defecating; note territorial neighbor in background. e) Rubbing forehead on neighbor's rump could be considered friendly, although the action deposits preorbital gland secretion and could thereby assert his territorial persona. f) Bull on left is inviting his opponent to urinate. g) He then analyses a sample by sucking it into his vomeronasal organ with the flehmen grimace. h) The two bulls kneel in combat position. i) One bull cavorts, demonstrating aggressiveness in a higher-intensity ritual. j) Solid horn contact, as both animals drop to their knees. k, l) Sometimes neighbors end or interrupt an encounter by lying down, both in this case. It is most likely to happen while kneeling and horning the grass, as kneeling also segues into lying. This sort of behavior supports the view of neighbors as "dear enemies."

FIGURE 8.6. *Continued*

Because aggression is largely ritualized, I termed these encounters challenge rituals. The duration of a ritual varied from less than a minute up to fifteen minutes; the average of seventy-one encounters timed from start to finish was six and a half minutes.

After observing a certain number of these encounters, I began to see a pattern in the recurrent actions. By the end of a year, the pattern seemed so clear that I was able to list the essential steps in a "typical" ritual in the order of their occurrence (fig. 8.6). It looked like quite a simple chain of actions and reactions, in roughly the following sequence:

a. Approach to bull standing in side-on presentation, head raised.

b. Both engage in displacement grazing, a neutral, unaggressive behavior.

c. One paws the ground, preparatory to defecating, which is not conspicuous in the gnu.

d. Bull in foreground is defecating. Note territorial neighbor in background.

e. Rubbing head on neighbor's rump, which could be considered friendly, although the action deposits preorbital gland secretion, which might be considered territorial.

f. Bull on left solicits neighbor to urinate.

g. He then analyses a sample by sucking it into his vomeronasal organ with the flehmen grimace.

h. One bull cavorts, demonstrating aggressiveness in a higher-intensity ritual.

i. Solid horn contact, as both animals drop to their knees.

j. Sometimes neighbors end or interrupt an encounter by lying down, both in this case.

When the time came to analyze the tape-recorded challenge rituals, it was soon apparent that this orderly sequence of events was largely illusory. Of over one hundred that I tape-recorded, no two protocols repeated the same actions in the same order. There is no such thing as a typical ritual. Nor is this really surprising, considering that there are at least thirty different actions that may be performed at any point, separately or in combination, by either participant.

One Protocol of One Actual Challenge Ritual

April 8, 1965, 15:55. A bull (A) is returning to his own territory. As he passes through a neighbor's territory (B), the latter walks rapidly after him, and both break into a run, then cavort—bucking, spinning, running sideways, shaking their heads, and kicking up their heels. In this fashion they enter A's territory, who now turns and comes back to meet B, stops, and defecates. Standing several yards apart, A urinates, then B. B stretches his nose under A's belly to smell the urine, then tilts head in *angle-horn* as he follows A's movements from the corner of his eye. A defecates while sniffing the urine, then performs *flehmen*. He circles behind B, who turns and circles with him in the *reverse-parallel position*. Their tails are held out slightly. Suddenly both whirl at once, drop to their knees in *combat position,* and make contact for an instant. Then they back off a yard on their knees and proceed to rub their heads and horns in the grass. A gets up and shakes his head, standing before B who continues horning. Now he also stands, and they perform the *alarm display:* gazing into the distance, stamping their forefeet, and giving alarm snorts (see sidebar). B shakes his head and lunges forward, at the same time elevating his tail. A comes to meet him, raising and switching his tail. Both drop to their knees and "wipe" the ground. B gives a *head-and-tail sweep* while still on his knees. A stands, gives a *head-and-tail sweep* and shakes his head. B gives a *head-and-tail sweep,* now also standing. A again shakes his head, then starts grazing. B shakes his head and walks away a few yards. Both graze, about 5 m apart, A moving deeper into his territory. B stays where he is, still grazing. A scratches his chin with a hind foot and resumes grazing, paying no further attention to B, who remains about 8 m away, near or inside their common boundary. Duration 3 minutes.

The alarm pause is a common feature of challenge rituals involving combat, as natural selection has eliminated individuals so absorbed in head-bashing that they fail to detect approaching predators, for example, lions. The alarm pause includes alarm snorts regardless of whether any predator is in sight. I suspect this behavior is subject to gamesmanship, as I have observed that one snorter sometimes takes advantage of his opponent's distraction to attack. It reminded me of boyhood wrestling matches when my opponent would suddenly point at me and say, "Look over there," then knuckle my arm twice if I fell for the ruse. "Two for flinching!"

Lying Down on the Job?

Occasionally one contestant, and rarely both, would lie down in the middle of a challenge ritual (see fig. 8.6). It looked absent-minded, as though the bull had forgotten all about his antagonist and, finding himself on his knees, simply completed the act of lying. In this case, A followed suit. It looked anything but aggressive—and more like companionship. Taken together with rubbing the muzzle on the opponent's rump or shoulder, these actions might be classified as sociable. I have seen bulls lie thus for some minutes, whether in their own or another's territory, and placidly ruminate within a few yards of an antagonist. Lying down usually put an end to the challenge ritual; the other bull would graze, call at passing cows, or perhaps walk over to challenge another neighbor. But sometimes the resting animal would get up and continue the encounter as though it had not been interrupted.

With no herd in his territory 80 percent of the time, the average bull leads a largely solitary existence.[6] The challenge ritual may well serve the need for social contact. One's neighbors may not be friends but could be treated as coequal rivals or "dear enemies."[10]

Lone bulls also spend time with other antelopes that happen to be near their territories. Is this another indication of their sociability? Or is it done to reduce exposure to predators? Or both?

The Male-Male Urine Test: An Identity Check?

In the normal course of events, flehmen (see fig. 8.6) is the response of males to the stimulus of female urine. Hormone breakdown products in the urine enables them to assay female reproductive status in the vomeronasal organ (as explained on p. 81). That urination by one male should routinely release flehmen in another male I had never heard of in any other bovid—except the black wildebeest.* The fact that urination-flehmen was seen in over 40 percent of the rituals indicates the importance of this behavior in the relations of territorial gnus. If a bull failed to urinate spontaneously during an encounter, his partner prompted him by extending his nose under his belly in 9 percent of the rituals. As the vomeronasal organ is adapted to the analysis of reproductive hormones in urine, testing a male's urine should reveal a dear enemy's

---

* I later learned from Fritz Walther (pers. comm.) that in zoos male eland and Gazella dorcas respond with flehmen to the urine of conspecific males.

testosterone level. The enactment of the sequence during the challenge ritual could thus serve as a chemical check of territorial credentials.

When Sedentary and Migratory Populations Overlap

Inevitably, the migratory population that follows the greenest pastures inside the Crater moves into an area occupied by a resident subpopulation. Suddenly, the established network of permanent territories and semiexclusive small herds is overrun by thousands of strangers.

What I found most interesting was the impact on the territorial network. When a wildebeest aggregation moved into a resident neighborhood, the distance between bulls was effectively halved, from 100 m to 50 m, as the accompanying outsiders infiltrated and set up shop between the resident bulls. The level of activity and noise rose and if a cow in estrus was present reached fever-pitch. Yet the compression of the network did not greatly raise the level of conflict, probably because all the extra bodies partially blocked territorial neighbors' view of one another.

Tolerance of extreme territorial compression is what enables the wildebeest's territorial system to function during the rut and migration (see chapter 10). When aggregations that moved into my study area moved on, life returned to normal. Judging from the known bulls and cows, most of the residents stayed put until pastures stopped growing at the end of the rains.

FEMALE LIFE HISTORY

In migratory populations, most females separate from their mothers as yearlings, like males, beginning at nine months. Yearlings become a separate class because of the attraction between individuals of the same age and stage of development. Territorial males segregate the sexes in the process of breaking up aggregations into small units averaging dozens of females and young.[1,11] They easily detect adolescent males in these pseudoherds. The ousted yearlings band together as a distinct class in bachelor herds. Meanwhile, territorial males herd yearling heifers together with the other cows and calves. Some heifers continue to associate with their mothers until the next calf is born. But in their efforts to maintain the maternal bond, interference with the mother and her calf provokes and aggressive response and rejection. However, the bond may continue indefinitely if a mother is barren. In Ngorongoro, I often saw a second-year female, branded in her first year, sticking with

FIGURE 8.7. Yearling following mother and newborn, chased by bull.

her barren mother. Moreover, should a newborn die, mothers will allow yearling offspring to suckle (more in chapter 11). So there is a potential payoff for yearlings that persist in following their mothers through the calving season, despite maternal rejection.

It is harder for yearlings to remain with their mothers in resident subpopulations. Here the territorial males act as gatekeepers. Any individual that is treated aggressively by the cows in a nursery herd is likely to be chased out by a territorial male. This applies to yearlings as well as to outsiders trying to join these semiexclusive groups. So yearlings that interfere with the mothers of new calves get booted out by resident territorial bulls (fig. 8.7). It can even happen when the mother tries to defend her yearling from the bull. That may change during the next rut if heifers have reached breeding age—which happens as early as sixteen months if they're sufficiently well grown. But four out of five heifers only come into season a year later. Unbred two-year-olds are often found in bachelor herds.

Thus small herds appear to be semiexclusive and enduring associations derived from the tendency of females with young calves to band together and probably also from continuing bonds between cows and

their female offspring. Indeed, as previously pointed out, it is characteristic of antelopes with polygynous mating systems that females can
remain in maternal herds, whereas the intolerance of breeding males
leads to separation of adolescent males. The impala is a well-studied
example.[12]

What happens in resident wildebeest herds shows that the process is
more complicated than in the mobile aggregated stage. A further com-
plication is the interchange of the two systems. In the Western Serengeti,
the presence of younger calves in resident herds belonging to immigrants is proof of interchange. In Ngorongoro, nursery herds join aggregations on the green belts as the dry season drags on. At first they com-
mute, returning in the evening to their regular range, but in time stay in
the aggregations and may lose their identity. But my observations of
herds containing known individuals showed that many of them became
resident again when rainfall rejuvenated their home ranges. Temporary
interchanges even take place overnight when resident herds decide to
spend the night with aggregations on the central plains, when shorter
grass and greater numbers make them safer than in isolated groups. The
resident bulls do what they can to prevent their passage, but all in vain.
Next morning the herds return home.

In my first report[11] on the resident nursery herds in my Ngorongoro
study area, I noted that eight herds of females and young observed 47
to 86 times from January through March were found within a 100-yard
radius of a fixed reference point two-thirds of the time (ranging from 30
to 94 percent), usually within the territories of one or two bulls. Herd
size ranged from 8 to 47, including 5–30 cows, 1–12 calves, and 0–6
yearlings, each herd with one bull.

Herd composition was not totally consistent but subject to daily and
even hourly variation. Nevertheless, the same number of cows, calves,
and yearlings were found together in 64 percent of all sightings. Inac-
curate counts would account for some discrepancies, especially in the
herd of 45–47. But more significant than variations in herd composition
and frequency in the same spot, and the absence of some herd members
for days on end, eventually the herds in my study area turned up again
in the same places with all members accounted for, except in a few cases
when a cow or a calf were missing for a long time and considered dead.
In my next report, through January 1964, I found that the average
group in a sample of 1,000 wildebeest in small herds numbered 9 animals. Small herds numbering 9–10 have been consistent in Ngorongoro
through succeeding decades. Most recently, 186 wildebeest in 22 herds

that I sampled on March 31, 1999, in my Munge Valley study area averaged 8.45 (range 2–27), and 96 animals in 10 herds I sampled on March 23, 2002, averaged 9.6 (range 2–30).

The season of resident small herds in Ngorongoro begins with the splitting up of dry-season concentrations at the beginning of the short rains in November, as pastures abandoned during the long dry season (May–October) are rejuvenated. The change is marked by a major peak in territorial activity, as former absentee landlords reclaim property and prospecting newcomers stake their claims. Small herds form before the calving season and consist mainly of pregnant cows, most or all of which calve in situ—not on the main calving grounds. Each herd acted largely independently and remained within a restricted home range, often less than 100 ha during the wetter half of the year (November–May), encompassing the territories of four or five resident territorial males. Captive migrating females kept in a 25 ha enclosure behaved in the manner of resident herds, including establishment of a dominance hierarchy (see chapter 10).[13]

Remarkable attachment to place *(Ortstreue)* within a patch as small as 100 sq. m suggested a feedback loop during the time of maximum grass growth. It seemed that small herds obtained optimum nutrition by mowing the same small lawns, a microcosm of the lawns maintained by aggregations on the short-grass plains (chapter 9).

REFERENCES

1. Watson 1969.
2. Mduma, Sinclair, and Hilborn 1999.
3. Georgiadis 1985.
4. Watson 1967.
5. Hopcraft et al. 2010.
6. Estes 1969.
7. Talbot and Talbot 1963b.
8. Estes 1967a.
9. Hediger 1949.
10. Wilson 1975.
11. Estes 1963.
12. Murray 1982.
13. Moss Clay 2007.

Cooperation and Competition among Twenty-Seven Herbivores That Coexist with the Wildebeest

The Serengeti ecosystem supports twenty-eight ungulates together with ten large carnivores that eat them. This is an extraordinarily rich and diverse ungulate community (fig. 9.1). How can all these species coexist with the wildebeest, which outnumbers them all put together and is so dominant that this one ruminant grazer defines the whole ecosystem? The impact on other grazers that share the savanna is particularly relevant.

"Under these circumstances," wrote Fryxell et al. in *Serengeti III*,[1] "one would normally expect to witness competitive exclusion by the dominant competitor and hence reduced diversity; yet unparalleled diversity is the name of the Serengeti game" (277). Surprisingly, considering the decades of ecological research carried out in the Serengeti, the explanation for this diversity is currently unknown. Only a few members of the ungulate community have been studied in sufficient depth to fit all the pieces into the puzzle of a species' ecological niche. Research has focused on the most numerous and dominant species: the migratory wildebeest, zebra, and Thomson's gazelle; and local populations of buffalo, topi, hartebeest, and impala. Elephants were studied after they invaded Serengeti NP in the late 1960s, but no researchers focused on this species from the mid-1970s until 2005.[2]

To be sure, enough is known of the ungulates' ecology to describe the major differences in their habitat and food preferences. The niche space

FIGURE 9.1 (TABLE)   THE 28 HERBIVORES IN THE SERENGETI ECOSYSTEM

Common Name	Latin Name	Population
Wildebeest	*Connochaetes taurinus*	1,086,754
Zebra	*Equus burchelli*	183, 815
Thomson's gazelle	*Gazella thomsonii*	328, 620
Grant's gazelle	*Gazella granti*	54, 628
Topi	*Damaliscus l. korrigum*	38,990
Coke's hartebeest	*Alcelaphus buselaphus*	16,043
Impala	*Aepyceros melampus*	90,692
Waterbuck	*Kobus defassa*	1,186
Warthog	*Phacochoerus africanus*	3,737
Buffalo	*Syncerus caffer*	30,276
Giraffe	*Giraffa camelopardalis*	10,460
Elephant	*Loxodonta africana*	2,360
Eland	*Tragelaphus (Taurotragus) oryx*	15,773
Roan	*Hippotragus equinus*	180
Oribi	*Ourebia ourebi*	7,000
Bohor reedbuck	*Redunca redunca*	
Mountain reedbuck	*Redunca fulvorufula*	200
Klipspringer	*Oreotragus oreotragus*	< 200
Dik-dik	*Madoqua kirkii*	
Steinbuck	*Raphicerus campestris*	
Grey duiker	*Sylvicapra grimmia*	
Bushbuck	*Tragelaphus scriptus*	43
Lesser kudu	*T. imberbis*	scarce
Greater kudu	*T. strepsiceros*	53
Oryx	*Oryx gazelle callotis*	100
Bushpig	*Potamochoerus porcus*	
Rhino	*Diceros bicornis*	13

SOURCE: Adapted from Sinclair, Packer, et al., eds., 2008, Appendix.

of at least a third of the twenty-eight listed ungulates are very restricted in the Serengeti ecosystem and more or less exclusive: the desert-adapted oryx, largely confined to the Salei Plain, the most arid part of the eco-system; the mountain reedbuck, specialized on hilltop grassland; the klipspringer's exclusive domain of kopjes and cliffs; greater kudu, largely confined to mountainous terrain in East Africa, found only in the Maswa Game Reserve, in the transition from savanna to *miombo* woodland; lesser kudu, arid bush, found only in the Loita Hills; roan, primarily a *miombo* antelope marginal in the Serengeti, mainly in the Maswa Game Reserve; bushbuck, a cover-dependent, edge species;

bushpig, wooded and bushed habitat, mainly found along western rivers, which eats roots, bulbs, and fallen fruit; warthog, diurnal grazer in tall grassland, underground refuge–dependent; hippo, confined to rivers bordered by grassland; black rhino, pure browser; giraffe, pure browser on trees.

ECOLOGICAL CATEGORIES

Let us consider factors still at the macrolevel that sort the ungulates into categories. Adaptation of the different tribes and subfamilies for particular environments were considered in more detail in chapter 1.

Ruminants versus Nonruminants

The adaptability of the ruminant digestive system has enabled many antelope species to evolve and partition the environment into much narrower segments than nonruminants.

Sorting by Size

Size is the most basic division among the ruminants, affecting diet, habitat choice, predation risk, and mobility.[3,4] Small ruminants like dik-dik and steenbok have a higher metabolism and depend on the most nutritious available vegetation. They are the so-called concentrate selectors. A pointed muzzle and narrow incisor row enable them to pick and choose the most digestible browse. Not having to cope with the more fibrous and abrasive grasses, the browsers have less complex digestive systems that enable them to feed and quickly digest leaves and forbs. Grazers like the reedbucks, waterbuck, topi, and wildebeest have bigger, more developed stomachs wherein ruminant bacteria extract nutrients from grass, slowly and completely. All the dominant ungulates are grazers or mixed feeders, grass being by far the most available food. The smallest grazer is the 20 kg oribi.

Cover-Dependent versus Cover-Independent

Cover-dependent species are distinguished by concealing coloration and other traits noted in chapter 1. Most are solitary or live in monogamous pairs. Conspicuous coloration and gregarious social organization are traits widespread among bovids adapted to open habitats.[5]

Diet and Social Organization

Selective browsing of foliage growing on separated plants and limited visibility favors individual over coordinated activity. Even sociable browsers live in much smaller herds than the grazers occupying open habitats.

Water-Dependent versus Water-Independent

Ungulates capable of meeting their water needs from their diet can afford to be sedentary.[6] Dik-dik, steenbok, suni and other dwarf antelopes, common duiker, and mountain reedbuck gain water by browsing green leaves and forbs. Gazelles, kudus, and eland can live in waterless habitats. Surprisingly, so can hartebeest and even wildebeest, considered water-dependent (D. Peterson pers. comm., 2010). Most of the water-independent species, including oryx and Thomson's gazelle, will go to water when available, but I've only once seen a Grant's gazelle drink (in well-watered Ngorongoro Crater). This gazelle and the oryx take advantage of their water-independence by occupying the short-grass plains during the dry season,[7] reversing the direction of the water-dependent species that follow the increasing rainfall gradient northwest and north once the rains end.

How Predation Affects Habitat Preferences

Predation has been shown to influence habitat choice and social organization, largely dependent on size.[8,9] Those above 150 kg are vulnerable only to the largest predators, notably lion, spotted hyena, and leopard. Small ungulates are preyed upon by the whole range of carnivores, from wildcats on up to the lion. Ranging of dwarf antelopes (along with other small herbivores and even small carnivores) that depend on hiding is restricted. Monogamous pairs defend exclusive home ranges as territories, intimate knowledge of which offers their best hope of surviving to reproduce. From this perspective, water-independence can be seen as an antipredator strategy. Nevertheless, predation controls their populations, especially through culling of dispersing offspring searching for vacancies in their preferred habitat.

Among the twenty-eight ungulates, medium and large species have far less to fear from the array of small predators. But they are the preferred prey of the top carnivores, with wildebeest and zebra heading the menu. The largest herbivores, starting with the buffalo and including giraffe, hippo, rhino, and elephant, are practically immune as adults. Ungulates weighing over 150 kg, especially the grazers and mixed feeders (gazelles, impala, topi, hartebeest, wildebeest, zebra, eland) that depend on flight and herd membership to survive, are free to move about large home ranges. Yet studies of their distribution via aerial observations and monitoring of GPS-collared wildebeest and zebra via satellite have shown that predation risk also constrains their movements. They avoid habitats that provide cover for ambushes by lions (leopards, too, for those in the 150 kg size range): closed woodland and tall grassland and steep, rocky, and wet terrain that impedes flight, drinking places in rivers (where crocodiles enter the picture), water-holes surrounded by vegetation, passageways through cover, and embankments that can conceal a lion. Speaking of which, an embankment created by excavating a ditch just south of and parallel to the main road through the Western Corridor, apparently intended to curtail flooding of the road during the rains, provides a wonderful amenity for all ambush predators on what used to be a level playing field.

Records kept over the past twenty years of lion kill sites confirm that the prey species have the very best reason to keep their distance from all potential ambush sites.[8,10] If you watch antelopes or zebras at waterholes or passing through patches of concealing vegetation, it is obvious that they are very aware of danger. Even a minor disturbance, like pigeons suddenly taking flight, can set off a stampede. Given a choice between a muddy, alkaline waterhole in open, short grassland and clean water in a nearby river that could harbor crocodiles, or vegetation that could hide a lion, the isolated waterhole will be the first choice every time.

Top-Down and Bottom-Up Population Control

Species whose abundance is limited by predation are said to be under top-down control. Probably most of the small and cover-dependent ungulates fall into this category. Species whose abundance is limited not by predation but by food supply are said to be subject to bottom-up control. These two different types of population control act simultaneously on the diverse Serengeti herbivore community.[3,4,11] Wildebeest, topi, buffalo, giraffe, hippo, rhino, and elephant are subject to bottom-up control. In the gnu's case, predation by about eight thousand lions and hyenas has hardly any impact on the immense migratory population. But predation may limit the relatively small resident populations.[12] One would suppose that a quarter million zebras would be subject only to bottom-up regulation, but their failure to increase in over four decades has caused Serengeti ecologists to suspect top-down control by predation,[13] possibly on foals.[8,14] If so, pack-hunting hyenas would be the prime suspects,[15] because the zebra's group defense by the family stallion and his mares must be an effective antipreda-tor strategy—how could it persist otherwise?[6] Absent observational evidence of widespread foal predation, this explanation is suspect.

However, Sinclair[16] offers a more plausible, though still circumstan-tial, explanation for top-down control of the zebra. He points out that predation accounts for 60–74 percent of annual adult zebra mortality, compared to only 25–30 percent of wildebeest and buffalo mortality. Moreover, half the zebra population does not migrate with the wilde-beest but remains scattered throughout the woodlands, where they are in effect resident prey and are killed by lions three times more frequently than would be predicted. Thus zebra are more available to predators throughout the year than are migratory wildebeest.

HABITAT DIVERSITY

Habitat diversity occurs at multiple levels and reflects multiple influ-ences. For instance, the gradient of increasing rainfall, which runs from southeast to northwest across the Serengeti ecosystem, is associated on the Serengeti plains with changes in the growth form of grasses from short mat-forming to medium and tall tufted species and in the wood-land zone with increasing tree height, density, and species richness.[17–19]

Physical features of the landscape interact with rainfall to break the gradient of vegetation response into more or less discrete units of simi-lar habitat. At the macro level the best example is the Serengeti plains, where a combination of relatively low rainfall (500–600 mm/year), fine-textured volcanic soils, and shallow hardpan favor grasses over woody plants.[20–22] The Serengeti plains are classified by Gerresheim[23] as a single land region defined as a land unit with a common geologic history and, often, a local climatic pattern. In contrast, the woodlands north and west of the plains have higher rainfall and occur on better-drained soils derived from Precambrian rocks.[22,24] At this scale of habitat definition, the wooded areas of the Serengeti ecosystem are shown to be more diverse than the plains in that they comprise eight land regions.[23] This division of the Serengeti ecosystem into two parts is of the greatest importance for the migratory wildebeest and zebra populations.

Landscape features also influence habitat diversity at the local level, where flows of surface and subsurface water differentially redistribute soil nutrients and soil particles, creating topographically related series of soil/vegetation communities (catenas) (fig. 9.2).[25] A typical example is a grassy hill with coarse, well-drained soil and short grasses on top,

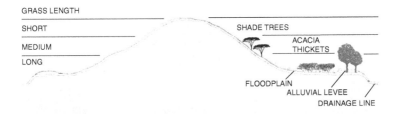

FIGURE 9.2. Characteristic topography of the Serengeti and much of the African savanna is a series of undulations, referred to as catena. The distribution of long and short grass is affected by the drainage from the elevations. Water runs off the elevations and into the depressions, so that the grass in the depressions tends to be longer than the grass on the elevations. In longer grass, some of the grazing animals have difficulty reaching the parts that are most nourishing for them. From Bell 1970, fig. 1.

finer soils with taller grasses on the slope, and heavy clay soils with impeded drainage at the bottom where tall grasses or sedges remain green well into the dry season. Different topographic forms produce different catenas, contributing further to habitat diversity.

Catenary grazing has an important role in the Serengeti. Analogous to the movement of grazers from the short-grass plains to the long grasses of the north, except on a smaller scale, it allows small herds of wildebeests and topi to remain sedentary throughout the year.[26,27] Catenas are widespread within the Serengeti NP, as suggested by the topographic relationships of many of forty-four woody species vegetation types mapped by Herlocker.[28] Indeed, the land systems used as the basic building blocks for the Serengeti Landscape Classification[23] are often based on catenas. (The different African savanna types are classified in chapter 2, and the variety of *Acacia* species in the Serengeti woodlands is discussed in chapter 6.)

At the smallest scale, even slight differences in topography on a seemingly featureless plain can alter soil texture and drainage enough to affect plant composition. Think of it as a microcosm of the catenary sequence on a grassy hill. For example, on Ngorongoro's central plain, bamboo grass *(Pennisetum)* grows in slight depressions, while star grass *(Cynodon)* occupies adjacent ground a few centimeters higher. Soil mineral content can also vary greatly with texture and drainage. In the Serengeti NP, soil samples taken in adjacent vegetation plots often show differences in composition as great as samples from different regions.[29]

Building on earlier efforts to classify and map Serengeti vegetation types,[24,28,30–34] a working group comprising ecologists and paleoecologists

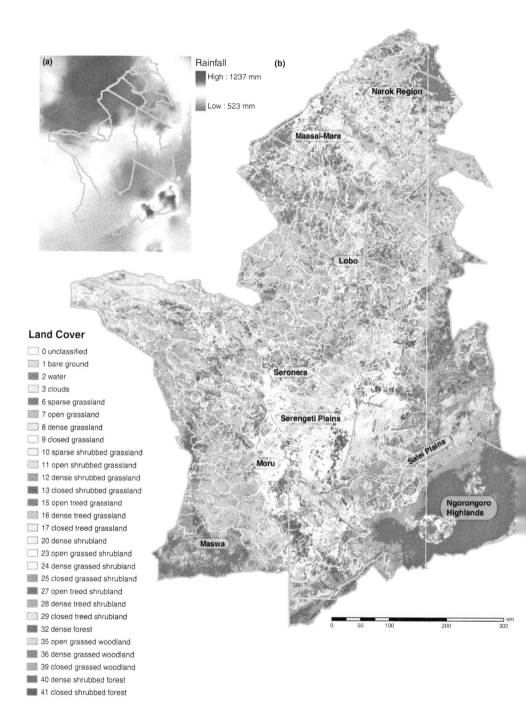

(a)

Rainfall

■ High : 1237 mm

□ Low : 523 mm

(b)

Narok Region

Maasai-Mara

Lobo

Seronera

Serengeti Plains

Salei Plains

Moru

Ngorongoro Highlands

Maswa

Land Cover

□ 0 unclassified
▨ 1 bare ground
■ 2 water
□ 3 clouds
■ 6 sparse grassland
▨ 7 open grassland
□ 8 dense grassland
□ 9 closed grassland
▨ 10 sparse shrubbed grassland
▨ 11 open shrubbed grassland
■ 12 dense shrubbed grassland
■ 13 closed shrubbed grassland
▨ 15 open treed grassland
▨ 16 dense treed grassland
▨ 17 closed treed grassland
■ 20 dense shrubland
□ 23 open grassed shrubland
□ 24 dense grassed shrubland
■ 25 closed grassed shrubland
▨ 27 open treed shrubland
▨ 28 dense treed shrubland
□ 29 closed treed shrubland
■ 32 dense forest
▨ 35 open grassed woodland
■ 36 dense grassed woodland
▨ 39 closed grassed woodland
■ 40 dense shrubbed forest
■ 41 closed shrubbed forest

0   50   100   200   300 km

involved in Serengeti research on interrelated topics collaborated to produce the first detailed vegetation map of the entire Serengeti ecosystem (map 9.1).[22] They used satellite photographs to delineate an intricate mosaic of vegetation units (patches) based on dominant plant life form rather than species composition. Each patch was placed into one of twenty-four land cover classes based on the pecentage cover of the primary and secondary life-forms—tree, shrub, and grass. These ranged from sparse grassland through open shrubbed grassland to closed shrubbed forest.

Reed and colleagues[22] then related the number, density, and diversity of the mapped vegetation patches to rainfall and topography. The results of their study showed that topographic heterogeneity contributes directly (through the influence of topography on the distribution and availability of water) to coexistence of woody and herbaceous patches across the landscape and that rainfall also influences vegetation characteristics. However, by themselves, rainfall and topography explain only a fraction of the total variance in patch characteristics and diversity.

For instance, from the Serengeti plains, which contain the largest continuous patches of homogeneous land cover in the Serengeti ecosystem, grassland patches become increasingly smaller along the gradient of increasing annual rainfall before becoming large again in the northern and westernmost parts of the Serengeti NP where rainfall is highest. Where edaphic factors, such as occur in the Serengeti plains, do not overwhelm its effect, rainfall controls the potential for woody cover at totals below 675 mm per year, but disturbance factors are more influential at higher rainfalls. Woodland (and overall vegetation) heterogeneity peaks at about the middle of the rainfall gradient (~890 mm/year). Thus rainfall and topography are not the only factors influencing

MAP 9.1. (On previous page) A structural vegetation map of the Serengeti ecosystem produced from satellite imagery using data from over 800 ground-truthing points. The differentiation of patches was based on the average patch area and distance to nearest neighbor determined for grassland, shrubland, and woodland vegetation types. The vegetation map includes a detailed summary of the land-cover classes and the diversity (heterogeneity) of the vegetation. The researchers found significant relationships between land cover diversity and soil moisture and also between diversity and average rainfall. The average area of grassland patches is significantly related to average and variation in rainfall and the soil moisture index. The vegetation map and analysis thereof resulted in three significant causal explanatory models that demonstrate that both rainfall and topography are important contributors to the distribution of woodlands and grasslands in the Serengeti. From Reed et al. 2008, fig. 2.

FIGURE 9.3. Aggregation of species on a possible hotspot on the Sabora Plain, Grumeti Game Reserve.

habitat diversity within the Serengeti ecosystem. Fire, herbivores, and ultimately humans are also important.

The large patches of tall grassland in the far north and west of the Serengeti NP may reflect burning and other human impacts.[35,36] On the other hand, exclusion of fire from Ngorongoro Crater changed the grazing ecology, promoting tall grasses over shorter species (see chapter 6).[37]

Fire and browsing by elephants greatly affect the composition and structure of woody vegetation in the park, including retarding the regeneration and replacement of fire-resistant tree species[17]. Varying intensities of fire and browsing by elephants since at least the 1890s have caused wide fluctuations in the relative amounts of grassland and woody vegetation in the Serengeti ecosystem, especially in the high rainfall areas of the north. Both factors ultimately reflect human activities, such as starting fires, poaching, and habitat destruction outside the park.[38] Grazing by herbivores strongly influences grassland biodiversity through its impact on life-forms, species composition, and mineral status, especially in the creation over decades of "hotspots," places where grazers of different species consistently congregate (fig. 9.3). Trampling of seedling trees and tree horning by wildebeest help to maintain open grassland.[38,39] Fencing almost immediately results in declines of diversity.[40-43]

THE MYSTERY OF HOTSPOTS

While driving the main road west from Seronera to Kirawira, you may travel miles without seeing many animals, but now and again you come upon a concentration of grazers, say, wildebeest, zebra, topi, Thomson's gazelle, and warthog. Or approaching Seronera driving through the taller grassland of the Serengeti plains, you pass a concentration of topi, hartebeest, and both gazelles in medium grassland. You may wonder what has brought them together on this spot. A shared preference for the grasses growing here? Safety in numbers? You may have read or been told that a mixture of species increases the chances of detecting danger through their differing sensory faculties. A much-cited example: giraffes use their great height to spot danger and warn associated smaller members; zebras or another species with great noses are better at detecting hidden lions or hyenas. This sounds plausible, but the evidence is almost all anecdotal, not based on scientific research.

It was McNaughton,[44] whose research on Serengeti soils and grasses spanned decades, who found that hotspots are different from adjacent grassland and are stable in space and time for at least two decades. They occur in regions receiving rainfall of more than 700 mm a year. The herbage has higher nutrient concentrations than in nearby grasslands used by few ungulates. Soils with concentrations of magnesium, sodium, and phosphorus, important for pregnant and lactating females, occur in concentrations that meet their dietary requirements. Moreover, experiments showed that soils in high-use areas had higher sodium and nitrogen content. Perhaps most important, grazing greatly increased sodium mineralization rates, which resulted in preferred sodium-rich leafy regrowth. Here is another prime example of herbivore cultivation: concentrated grazing increases green-leaf concentrations, nitrogen cycling, and sward density and bulk, thereby increasing ungulate carrying capacity.[44-48]

The cooperative study that produced the map that shows the habitat diversity of the Serengeti ecosystem in unprecedented detail (see map. 9.1)[22] enabled Anderson[43] to pick out probable hotspots. (Anderson was second author of the published map.) The hypothesis that guided their research was that the distribution of Serengeti's resident herbivores was determined simultaneously by top-down and bottom-up processes. That is to say, the patches where herbivores gathered afforded the most nutritious food resources at a safe distance from vegetation and terrain that made them vulnerable to predators, especially the lions that live in the woodland zone.[11] Study objectives were twofold: to use remote sensing and GIS to predict the spatial distribution of hotspots across the ecosystem and to investigate the ecological processes that determine hotspot distribution, emphasizing resource- and predation-related factors.

Attention focused on the seven grazers called the "hotspot species": zebra, wildebeest, topi, hartebeest, gazelles, and warthog. The regions surveyed were north and northwest of the Serengeti Plain, where annual rainfall was more than 650 mm, as McNaughton's research had established that permanent hotspots were rare below this threshold.

Anderson et al.[43] found that hotspots in the Serengeti typically occur in areas that are relatively flat and away from rivers and concealing vegetation, that is, locations where lions have most hunting success. Repeated grazing and fertilization of hotspots maintain them as grazing lawns.

In contrast to McNaughton,[44,48] Anderson et al.[43] found soil fertility to have a positive direct effect on leaf Na and Mg. Topography affected soil wetness by influencing plant biomass; shorter grasses were associated with flat terrain. Shorter grasses are more nutritious, and the concentration of sodium in the leaves is particularly attractive to Serengeti grazers. Still unknown is whether these influences are mediated by changes in plant species composition. Either way, landscape features have multiple important influences on hotspots from the standpoint of both predation risk and nutrient availability.

When all was said and done, the maps created by the computer models predicted less than thirty Serengeti hotspots, of which perhaps five are 10 sq. km or larger. How can there be so few in an ecosystem of some 25,000 sq. km? It just doesn't compute. According to McNaughton and Banyikwa,[42] the resident herds of topi, buffalo, hartebeest, impala, waterbuck, reedbuck, roan, oribi, and gazelles, which represent the principal reservoir of mammal diversity in the ecosystem, "are not uniformly distributed in space. Instead they are concentrated in localized 'hot spots'" (64).

Still more surprising, how and why hotspots originate remains unknown. How explain that they are fixed in space and time measured in decades? A possible clue: abandoned Masai cattle bomas remain localized hotspots for many years, thanks to the accumulation of urine and dung in the cattle kraals. That Serengeti NP hotspots might have originated from earlier pastoral occupation has been discounted, because rinderpest and dense woodland made the region unsuited for livestock following the great rinderpest pandemic.[14] But Reid[49] suggests that pastoral settlements dating back centuries could have laid the foundation for hotspots that wild herbivores have maintained ever since. Indeed, recent archeological research in the Loita-Mara area suggests that pastoral livestock production was the first form of food production in the Serengeti ecosystem and other regions of East Africa with bimodal rainy seasons dating back at least 2000 BP.[50] There is no reason to doubt that a succession of pastoralist tribes shared the savanna with the wildlife, as did the Masai who succeeded them in the past few centuries. Never mind that the rinderpest pandemic led to abandonment of the region in the 1890s; cattle and small stock had lived there for many centuries. An excavation of a pastoral settlement in Kenya hundreds of years old made the surprising finding that the earth in the subterranean cattle kraal still had a high nitrogen content.[50] That increases the likelihood that abandoned settlements were the loci of the present Serengeti hotspots.

In any case, it follows that intensive grazing by the mixed herds serves to cultivate hotspots through feedbacks, leading to swards of short grasses that accumulate sodium and phosphorus in their leaves, just as occurs on the Serengeti plains during the rains. Although soils in nonhotspots have the same minerals,[42] the difference is that sodium and phosphorus accumulate in the leaves of grasses on hotspots. The herbivores are cultivating grasses that accumulate these minerals along with nitrogen. Although the hotspots occur in regions with 700 mm or more of rainfall, they resemble the grazing lawns of the short-grass plains. Thus through coevolution of grazers and their preferred grasses, resident populations acquire the same nutrients vital to females during pregnancy and for development of their offspring as the migrants that congregate on the Serengeti plains during the rains.

One might suppose that the great habitat diversity (heterogeneity) of the Serengeti ecosystem would enable all the different herbivores to claim an exclusive set of species-specific ecological niches. Yet Fryxell et al.[1] could write, "Second, neither migration, differences in body size, nor adaptive patterns of landscape use appear sufficient to allow coexistence of even three grazers, let alone 28 species" (296). Nor, they add, is seasonal variation in resources, mediated by rainfall, adequate to explain the coexistence of all these species. Yet these species do coexist. That their niches cannot be described on the basis of current knowledge simply reflects ignorance of the full range of adaptations by most of the species for a particular slot in the resource continuum.

Moreover, as Fryxell et al.[1] note, "We still do not have a very clear idea about the biological features that permit coexistence in a speciose, yet similar, guild of terrestrial herbivores. It may be that we have failed to capture the crucial spatial scale at which these interactions take place" (297). It is also possible that exclusive use of resources by each species enables them to coexist, for instance, by selecting different parts of the same plants or specializing for different species or types of vegetation, starting with the separation between grazers and browsers.[1]

My own vision of how all these ungulates manage to coexist combines all spatial scales, from macro to micro, nested one inside the other like Chinese boxes. To illustrate, wildebeest and zebras range across the entire ecosystem, following the rainfall gradient from semiarid to well-watered savanna and back. Topi and buffalo that live in large aggregations have very large home ranges within a region, for example, in the Western Corridor. They move considerable distances between preferred wet- and dry-season pastures. But topi and wildebeest distributed in the sedentary-dispersed pattern (small herds) live in much smaller home ranges wherein they exploit catenas that contain the grasses preferred in wet and dry seasons within an area as small as a few square kilometers. The hartebeest is an outstanding example, as discussed in chapter 3. Then there are mosaic grasslands, such as those on the Ngorongoro Crater floor, that offer a smorgasbord of grasses and herbs within meters of one another. A striking example was the topi of Uganda's Queen Elizabeth National Park that moved about the Ishasha Plain from grassland patch to patch in aggregations up to a thousand strong.[6,51] At the smallest scale, the smallest ungulates have to seek and stay within very restricted habitats that offer security and a browse menu for all seasons.

Returning now to the original question, how can twenty-eight herbivores coexist when one is so dominant that it defines the ecosystem? Specifically, how does the wildebeest affect the other members of the community? The attributes that make this one antelope so perfectly adapted to this particular set of geologic, climatic, and ecological conditions were noted in chapter 6: size, conformation, feeding adaptations, migratory habits, and unique reproductive system. Depending on the habitat and the season, the wildebeest's effect on associated species can be positive or negative.

COMPETITION VERSUS FACILITATION

Setting aside the antipredator benefits of associating in multispecies herds, the members of a grazing assemblage can mutually benefit through their different feeding preferences. Arsenault and Owen-Smith[52] outline three ways in which the associated grazers stand to benefit: feeding facilitation, habitat facilitation, and population facilitation.

Feeding Facilitation among Grazers

The best-known example of feeding and habitat facilitation is the grazing succession first described by Vesey-FitzGerald[53,54] in Tanzania's Lake Rukwa floodplain and later by Bell[55] in the Serengeti plains. Species A, a bulk grazer of tall grasses, makes more grass accessible to species B by reducing grass height and removing stems, and species B crops the sward to the short stage preferred by species C. Buffalo, zebra, wildebeest, and Thompson's gazelle, the dominant grazers in the Serengeti ecosystem, are the exemplars of such a succession.

The reality of the Serengeti grazing succession was later called into question.[56–60] For one thing, wildebeest did not depend on zebras to prepare their pasture; they often took different migratory routes. Zebra also enter woodland and tall grassland that wildebeest avoid. Nevertheless, grazing successions do occur under favorable conditions.

Ngorongoro Crater is a classic example. A grazing succession can be clearly seen during the dry season, when the dominant grazers move onto the greenbelts bordering the swamps (fig. 9.4). Buffalo and cattle freely enter marshes and swamps while they are still flooded. Zebra are also bulk feeders on tall, stemmy grasses and will enter wetlands avoided by wildebeest. When cropping and trampling have reduced the tall vegetation to below knee height and the ground is nearly dry, the wildebeest move in to graze the more palatable understory grasses. After they have reduced the pasture to a close-cropped sward and moved farther into the shrinking greenbelts, Thomson's gazelle harvest grasses and forbs too short and sparse to sustain the bulk feeders. Herds of Grant's gazelle also join in, feeding mainly on forbs, herbs, and small shrubs. The last patches of green pasture are the drying marshes and swamps, cleared and mulched of reeds and sedges (assisted by swamp-feeding elephants) to be replaced by lush new growth crowded with Thomson's gazelles.

Monthly distribution maps that I prepared during the 1964 dry season showed that the percentage of the Crater wildebeest and zebra populations pastured on the greenbelts rose from nil in early May to 31 percent and 33 percent, respectively, in early July, to a peak of 42 percent and 60 percent in October, dropping to 37 percent and 50 percent early in November, following the onset of the short rains.[61] Yet the

FIGURE 9.4. Wildebeest and zebra associated on the greenbelt bordering Mandusi Swamp.

monthly distribution mapping I carried out in 1964–65 and 1996–99 showed that wildebeest and zebra populations overlapped more often than not only in the wet and early dry season. Maximum separation occurred in the late dry season, reflecting the zebra's wider habitat tolerance.[62] The same tendency to minimal association of the Serengeti populations in the late dry season was noted by Sinclair.[9]

Habitat Facilitation

While simply eating down and trampling tall vegetation may be the most obvious benefit of the Ngorongoro grazing succession, the improved nutritional quality of the regrowth has been noted (see chapter 6).[45,46,52,54] Grazing by armies of wildebeest in dense concentrations has an impact on the savanna wherever the migrants go, to the benefit of all during the growing season. Indeed, McNaughton[45] thinks the enhanced nutritional value of the new growth on grazed pastures may partly account for the abundance of wildebeest. Mowing the same pastures over and over again, they in effect farm the savanna in a vast rotational grazing system. Call it self-facilitation.[52]

But self-facilitation is operational only during the growing season. During the long dry season when the wildebeest have to feed on the tall grasslands of the woodland zone and the Masai-Mara, they lose condition, particularly lactating cows.[11,63] Then they also compete with rather than facilitate the foraging and/or habitat preferences of the long-grass grazers, even including elephants in the Masai Mara, as noted below.

Population Facilitation

Herbivores can benefit through population facilitation.[52,64] Access to nutritious herbage during the growing season promotes the reproductive success of grazing assemblages, enabling disadvantaged ungulates to survive exploitative competition during the season of want. Arsenault and Owen-Smith[52] consider this trade-off an important mechanism promoting coexistence of the grazers.

Mitigation by Migration

Migration not only enables wildebeest and zebra to utilize the grazing over a huge range of savanna habitats. It also mitigates competition with resident herbivores because the migrants move on without consuming all the forage they depend on to survive the dry season.

Mitigation by Fire

Both planned and unplanned fires sweep across the Serengeti every dry season beginning in June. The postburn flush that comes up in soil with sufficient moisture provides excellent nutrition that attracts all grazers. As fires caused by lightning have never been recorded in the Serengeti, the most nutritious dry-season pastures are man-made. The more rain that falls, the taller the grass and the more widespread the burns; fire cannot propagate on grassland cropped to a short sward.

Burning must date back at least several hundred years if not millennia had the Cushitic and other pastoralists who preceded the Masai conducted burns, and fire frequency increased tremendously with the worldwide distribution of matches early in the twentieth century (see Google History of Matches). The dependence of an antelope on burns is well illustrated in the case of the giant sable *(Hippotragus niger variani)* of Angola, whose dry-season distribution is closely correlated with postburn flushes.[65]

Parks managers and researchers have mapped the location and extent of Serengeti fires intermittently during aerial surveys as far back as the 1960s. In the early days fires burned over 80 percent of the ecosystem every dry season, but their extent declined to low levels by the 1980s. In fact, grazing by the Serengeti wildebeest as its numbers peaked in the 1970s removed so much tall grassland that fire frequency was reduced to the point that woodland was able to reclaim lost ground.[66] Yearly burns covered less than 25 percent of the ecosystem into the 1990s, then increased under the policy of early burning by park managers.[18,67,68]

So far in the present century, widespread fires start as soon as the grass and litter will propagate flames. As noted in chapter 6, the early burning policy in the Serengeti NP, intended to create a mosaic of burned and unburned savanna, is uncontrolled despite elaborate fire management plans and removes extensive pastureland wildebeest, zebra, topi, and buffalo would otherwise consume (author's observations).

HOW A BEHAVIORAL TRAIT SERVES TO MAINTAIN THE WILDEBEEST'S PREFERRED OPEN HABITAT

Wildebeest have the habit of rubbing their preorbital glands on the branches and trunks of trees, usually as the prelude to horning bouts. Olfactory communication is involved, as marking and horning are directed especially at sites that have already been marked and horned; and bouts begin with sniffing of the site. The behavior is also infectious, as wildebeest will line up and take turns rubbing and thrashing the same tree. However, vegetation horning is primarily an activity of adult males; other wildebeest horn much less often and vigorously. Bouts may last from fifteen seconds or less to as long as five minutes (fig. 9.5). It is clearly aggressive in character and may either be directed, as in encounters between territorial bulls, or an undirected display of aggressive mood.

What is extraordinary and generally overlooked is the environmental impact of wildebeest horning in large populations. Data collected on the incidence and effects of horning in 1979, 1980–81, 1986, 1998, and 2003 show that on the Serengeti Plain and in tree savanna of the woodland zone two out of three young trees of suitable size and accessibility have been horned, many repeatedly over a period of years. In numerous cases, the stems and branches of trees (especially *Acacia* spp. and *Balanites aegyptica*) have been destroyed every time they produced stems of thrashable size (fig. 9.6). The resulting growth in the case of *A. tortilis,* a dominant tree, is a supine hedge, within which a main stem of up to

FIGURE 9.5. A bull horning a small acacia, expressing a low level of aggression—behavior performed by all adult males and some females.

10 cm diameter may be found. The evidence supports the conclusion that destruction of woody vegetation by wildebeest was second only to fire and elephants in transforming Serengeti savanna woodlands into tree grassland during the 1970s and 1980s.[39]

IMPACT OF THE WILDEBEEST ON SELECTED UNGULATES

The effects of the migratory wildebeest on other ungulates can be negative or positive, depending on whether they are adapted to long or short grassland and closed or open habitats.

Thomson's Gazelle

Tommy that live on the long-grass plains in the woodland zone concentrate on the shortest patches. The rolling plains of the Grumeti Reserve afford good examples.

Tommy concentrate on a hilltop close to the Nyasarori ranger post, especially during the rains . It is the top of the catena, where the soil is coarse and well drained and short grasses prevail. The wildebeest migration directly affects the gazelle's dry-season distribution, as tommy concentrate on the grazing lawns where wildebeest concentrations have removed the tall grass. This is a clearcut case of feeding facilitation. Zebra may or may not be involved in the succession.

FIGURE 9.6. A sample of the damage to woody vegetation by horning, which is performed year-round, with a peak during the rut.

FIGURE 9.7. Thomson's gazelle concentration on the Serengeti Plain near the Seronera River.

The main tommy population that inhabits the Serengeti short-grass plains during the rainy season concentrates on the most nutritious pastures, including lawns mowed and fertilized by the returned wildebeest migration (fig. 9.7). Equally important, tommy exploit random variation in the amount of rain that local thunderstorms deliver by moving between patches that receive the most soaking and produce the lushest growth.[1] Measurements of lawns where tommy were grazing showed that they enjoyed the highest energetic returns on swards yielding 21 g per sq. m, comparable to production on a freshly mown lawn. The researchers concluded that the variable, localized rainfall regime is so important to the half million tommy that the population would be unlikely to persist on a savanna landscape of less than 2,500 sq. km. Far from being random, gazelle movements track the shifting location of patches that are most nutritionally and energetically rewarding.[1,45,69–71]

In the dry season the main population moves west up the rainfall gradient into medium to high grassland at the edge of the woodland zone. Here, too, tommy distribute themselves on the shortest pastures, often in association with topi, hartebeest, and Grant's gazelle. But the park burning program of the grassland on alternate sides of the main Seronera–Naabi Hill road, on a yearly or two-yearly schedule, has by far the greatest impact on the distribution of the grazers on the long-grass plains community.

Oribi and Reedbuck

Both oribi and reedbuck depend on tall grassland as their main anti-predator strategy. Bohor reedbuck live in the high grass and reeds lining the watercourses that cross and border the Serengeti plains, venturing into the shorter grassland nearby to graze. When fire or concentrations of the more abundant plains game remove their cover, the reedbuck band together in temporary herds. Oribi, which are common only in the higher rainfall areas of the woodland zone, also form herds when their cover is removed.

Mini-Browsers

It might seem that all the cover-dependent ungulates whose territories are within range of grazing aggregations are vulnerable to exploitative competition, but research on a Kenya ranch, using fenced plots to exclude browsers, showed that dik-diks played a major role in reducing recruitment of the woody plants they browsed. Although they only affected shrubs within their foraging height (< 0.5 m), reductions in the growth rate of twigs amounted to six times less recruitment to the next height class (0.5–1.0 m) within just three years. Combined with measured rates of shrub mortality in larger height classes, browsing reduced the rate of increase in shrub density nearly to zero. In short, browsers improved the pasture for grazers by suppressing woody growth.[72]

Topi

The detailed comparison of wildebeest, topi, and hartebeest feeding behavior in chapter 3 showed that wildebeest are best equipped to harvest swards of short, leafy grass; topi have the advantage in medium grasslands because of their efficiency at selecting green leaf; and hartebeest can meet their lower energy needs by gleaning the least fibrous leaves and shoots in tall, mature grassland.

During the dry season, wildebeest deplete the topi's food supply when large concentrations mow the pastures that enable them to maintain fair to good condition. The resident topi then have to move in search of the grassland they prefer.[27] My first experience with what could be called competitive exclusion of topi by wildebeest, on a small scale, was in May 1979 when an estimated quarter million gnus migrating off the short-grass plains spent nearly a week grazing the still-green tall grassland around Seronera. Several small topi herds that I was monitoring every day or two vanished from their accustomed home ranges soon after wildebeest had reduced their pastures to a short sward. Days later, I found two of the herds a dozen kilometers away in substandard habitat normally unoccupied by topi.

FIGURE 9.8. Topi and wildebeest on short grassland in the Grumeti Reserve. Wildebeest take bigger bites of shorter pasture than topi.

A striking example of displacement on a larger scale was an eye-opener for me in June 2011 on the Sabora Plain in the Grumeti Game Reserve. A sward of short grass several hundred hectares in size made a green patch in the surrounding tall grassland. It had been burned in April. The post-burn flush attracted a good many of the several thousand topi resident on the Sabora Plain. Their grazing, together with that of hundreds of zebras and Thomson's gazelles, kept the burn as short as a mowed lawn, which was gradually turning tan as regrowth slowed. I was surprised that topi could harvest enough grass to meet their needs and still more surprised when waves of wildebeest passing through from the Western Corridor began moving onto the burn (fig. 9.8). As they moved in, the topi moved out, heading northwest into tall grassland. The wildebeest proceeded to crop grass the topi had already grazed, demonstrating the advantages over other grazers of the broad incisor row, flexible lips, and rapid bite rate that make the wildebeest the premier bulk feeder of fresh grass.[27,73,74]

Topi reach highest density on floodplains where green grass is available year-round. They generally do not occur in regions with annual rainfall of < 740 mm. The Serengeti is the only place in East Africa where topi and wildebeest overlap, and the Western Corridor is their stronghold.[27] Topi concentrations of up to 9,000 on the corridor plains are very impressive, yet wildebeest outnumber them by 20 to 1.[75,76] As there is no evidence of ecological separation because of preferences for different kinds of grass, it is fortunate for the topi that the resident wildebeest population is less than 20,000 and that the migrants do not stay long enough to consume all the tall grass.[27]

Buffalo

Buffalo may be still more vulnerable than topi to wildebeest competition, as they cannot feed efficiently on grass too short for their automatic tongue action to bundle grass into big bites. Major die-offs of Masai Mara and Ngorongoro Crater buffalo populations occurred in 1993 and 2000, respectively, during severe droughts when all the grass was eaten.[38,62] Buffalo can browse to a limited extent to compensate for sparse grazing. Like the topi, this water-dependent bovine reaches highest densities in the higher rainfall areas of the ecosystem and prefers floodplains and swamps. Subpopulations live in traditional and exclusive home ranges. In the dry season, clans within the big herds separate into smaller herds that aggregate on remaining patches of preferred long grassland.[77]

Elephant

As grass is a prominent component of the elephant's diet, removal of the tall grasses in the Masai Mara by the wildebeest migration forces the elephant population to depend more on woody vegetation, increasing destruction of the woodland lining the Mara River and other watercourses.[38]

FURTHER EVIDENCE OF COMPETITION

Various studies of the ecological separation of the grazing ungulates agree that competition is reduced primarily through scaling in body size and secondarily through adaptations in feeding morphology, leading to different preferences for grass height and forage quality.[52,55,57,60,64,78–80] If, as evinced by a comparative study of buffalo, wildebeest, hartebeest, topi, and Thomson's gazelle diets in the Masai Mara, the grazers all choose the same grasses rather than diversify by choosing different species,[81] competition during the dormant season should make competition inevitable.

That Arsenault and Owen-Smith[52] "found no evidence that feeding facilitation has anywhere been translated into an increase in population abundance" (316) calls into question the widely accepted theory that facilitation is what enables the members of grazing communities to coexist. Instead of increasing along with the wildebeest population during the 1970s, both buffalo and Thomson's gazelle populations have shown a consistent downward trend since wildebeest numbers leveled off at over one million.[56,82] The zebra population has remained at about a quarter million over the past five decades.[1,83]

Buffalo numbers, in fact, did increase following the elimination of rinderpest in 1962, peaking at about 60,000 in the 1970s.[66] But their population has since been cut in half, mainly due to poaching (table 9.1). Their near-elimination from northern and western areas of Serengeti NP is linked with an increase in the numbers of topi, impala, and perhaps oribi, suggesting that competition with buffalo previously held them back.[9,14] Or was it because poaching also reduced the predator population in the northern Serengeti during the same period?[16]

Grant's gazelle also increased as the wildebeest population grew in the 1970s, consuming so much long grass that reduced burning allowed more woody growth favored by this browser-grazer to survive.[71] Similarly, *granti* benefit from scrub that invades overgrazed short grassland.

CAN TWENTY-EIGHT UNGULATES CONTINUE TO COEXIST?

Failure to increase and decrease of some key species raises questions about their continuing coexistence over the long term. Fryxell et al.[1] find these patterns consistent with the competitive exclusion hypothesis. However, given the limited time frame of Serengeti research, nothing is certain. As Fryxell et al.[1] write, "The spatially realistic models suggest it might take decades, if not centuries, for competitive exclusion to occur" (296). Also, basic questions remain unanswered, such as whether wildebeest have a competitive relationship with zebra that would explain their failure to increase or only a neutral relationship that begs the question whether top-down control through predation on foals (undocumented) is a likely explanation. Indeed, as noted earlier, there is some evidence that zebras may benefit from association with wildebeest because wildebeest, being most numerous, are more likely to be prey.[84]

As for the other ungulates that manage to coexist in the Serengeti ecosystem, understanding the uniqueness of each species' niche will require much closer study at a finer scale of habitat and dietary differences. The research on heterogeneity in the Serengeti ecosystem did not even attempt to model selective grazing for particular plant parts, despite strong evidence of species-specific differences,[85] because including such detail in the spatially realistic models would have considerably slowed the already complicated simulations.

Despite all the research that has been carried out in the Serengeti, the realization that ecologists are only beginning to comprehend the biological complexity of the ecosystem is sobering indeed. The importance of undertaking ever more detailed research, especially in view of the looming threats to this last intact African savanna ecosystem, cannot be overstated. (See chapter 12.)

REFERENCES

1. Fryxell et al. 2008.
2. Estes et al. 2012.
3. Mduma, Sinclair, and Hilborn 1999.
4. Sinclair, Mduma, and Brashares 2003.
5. Estes 1974.
6. Estes 1991a.
7. Walther 1972.
8. Hopcraft et al. 2010.
9. Sinclair 1985.
10. Packer et al. 2005.
11. Hopcraft 2010.
12. Ndibalema 2009.
13. Grange and Andrews 2006.
14. Sinclair 1995a.
15. Kruuk 1972.
16. Sinclair 2000.
17. Herlocker 1976.
18. Norton-Griffiths 1979.
19. Metzger 2002.
20. Anderson and Talbot 1965.
21. de Wit 1978.
22. Reed et al. 2009.
23. Gerresheim 1974.
24. Jager 1982.
25. Milne 1935.
26. Bell 1971.
27. Duncan 1975.
28. Herlocker 1975.
29. Anderson et al. 2008.
30. Gerresheim 1973.
31. Herlocker and Dirschl 1972.
32. Schmidt 1975.
33. de Wit and Jeronimus 1977.
34. Epp 1980.
35. Dempewolf et al. 2007.
36. Serneels and Lambin 2001a.
37. Amiyo 2006.
38. Dublin 1995.
39. Estes, Raghunathan, and Van Vleck 2008.
40. McNaughton 1979.

41. McNaughton 1983.
42. McNaughton and Banyikwa 1995.
43. Anderson et al. 2010.
44. McNaughton 1988.
45. McNaughton 1984.
46. McNaughton 1985.
47. McNaughton 1990.
48. McNaughton, Banyikwa, and McNaughton 1997.
49. Reid 2012.
50. Marshall 1990.
51. Jewell 1972.
52. Arsenault and Owen-Smith 2002.
53. Vesey-FitzGerald 1960.
54. Vesey-FitzGerald 1974.
55. Bell 1970.
56. Sinclair and Norton-Griffiths 1982.
57. Illius and Gordon 1987.
58. De Boer and Prins 1990.
59. Putnam 1996a.
60. Prins and Olff 1998.
61. Estes and Small 1981.
62. Estes, Atwood, and Estes 2006.
63. Watson 1967.
64. Owen-Smith 1989.
65. Estes and Estes 1974.
66. Sinclair 1979a.
67. Dublin 1986.
68. Sinclair et al. 2007.
69. Fryxell 1991.
70. Wilmshurst, Fryxell, and Colucci 1999.
71. Fryxell et al. 2005.
72. Augustine and McNaughton 2004.
73. Hofmann 1973.
74. Murray 1995.
75. Sinclair 1972.
76. Sinclair 1973.
77. Sinclair 1977a.
78. Jarman 1974.
79. Demment and Van Soest 1985.
80. Owen-Smith 1985.
81. Hansen, Mugambi, and Bauni 1985.
82. Dublin et al. 1990.
83. Grange et al. 2004.
84. Caro and Fitzgibbon 1992.
85. Jarman and Sinclair 1979.

CHAPTER 10

The Amazing Migration and Rut of the Serengeti Wildebeest

The rut of the Serengeti wildebeest is like nothing else on earth. Some half a million cows are bred during three weeks of frenzied activity. Yet surprisingly little of the activity actually involves mating. Most of it involves herding, chasing, and fighting among the bulls that are currently holding territories. They herd and chase the cows and fight one another as they compete to collect and hold a group. Success depends not only on their efforts, but on the willingness of the cows to be detained. When there is a general movement, the bulls' efforts are largely futile. They move back and forth in the passing parade in their distinctive rocking canter with heads high, calling continuously, but the horde keeps going (fig. 10.1). They're like small boats breasting a flowing river.

When an army of wildebeest numbering in the thousands stops moving, the bulls have the opportunity to round up and hold cows on their territories. The vanguard and also the tail end of the migration consists mainly of bulls. Those in the vanguard include bulls intent on staking out territories wherein they can detain passing females. Being out front enables them to meet most of the oncoming horde. How many they manage to keep, if any, depends on circumstances like the time of day, weather conditions, the location and amenities of the territory—good pasture, proximity to concealing vegetation, water, shade—and on the presence or absence of other cows.

How does a lone bull acquire a herd? In the territorial network encompassing every aggregation, most bulls are alone on their property.

FIGURE 10.1. A bull cruising beside a moving aggregation in the trademark rocking canter.

For example, of 513 territorial males I sampled during the 1980 Serengeti rut, only 116 had females on their territory = 1 out of 5. A bull canters toward an approaching cow, grunting a welcome as he tries to stop her inside his property. But blocking a cow's path doesn't work if she wants to move on, even if he chases her flat out. The chances of recruiting passersby improve if a group of cows and calves is already present. Having observed that cows and their accompanying offspring relay from herd to herd, despite the best efforts of the single territorial males to stop them, I eventually realized why most of the competition among territorial males is over having a herd. One needs a herd to attract passing cows, including, sooner or later, cows in estrus. You could almost liken the successful bull to a duck hunter deploying a spread of decoys to lure the desired quarry—a female in heat—to land in his spread.

Aerial photos of the clustered groups typical of the rut carried out in 1963, 1965, and 1966 recorded an average of 28.4 head ± 19.8 head (standard error of the mean) in a sample of 136 groups totaling 3959 animals.[1] The average of 4,032 animals in 353 Serengeti herds I counted during the 1997 rut was 11.4. The difference between 2,514 animals in 199 herds, 12.6, counted in late November 2006, was insignificant. In the Crater, during 1963–65 ruts, the average complement of 190 herds with territorial males was 17 (fig. 10.2). By comparison, herd size postrut and calving seasons of 1963–65 was 10.37 animals in 525 herds. The average number in 48 herds I counted during the 1964 Serengeti rut was 15.8, while the mean of 46 herds in the Ol Balbal in November

FIGURE 10.2. Bull keeping cows and calves in a tight group. Such clusters are typical of the rutting peak. The herd bull (head up) will do what it takes to keep the lone gnu in the background from horning in.

1964 was 18.7. The difference in the number of females and young in company with territorial bulls doesn't look that substantial. What brings the average herd size down is the presence of many small groups: 1–3, not counting the bull.

The chances of collecting and holding a herd depend importantly on the time of day. Efforts are least successful during the early morning and late afternoon feeding peaks, for obvious reasons. The best times are the transitions from feeding and moving to rest and rumination periods, from 8 or 9 A.M. to, say, 5 P.M., depending on cloud cover and ambient temperature. The animals remain active longer during cool weather. On sunny days bulls under trees have a major advantage over bulls in the sun, as illustrated in figure 10.3. On July 1, 1999, I counted 1,026 wildebeest in 63 herds between 10:30 A.M. and 4:00 P.M.; 246 were with territorial bulls in open grassland, and 780 were in shaded territories. Average herd size was 3.9 versus 12.4 in the shade. At the risk of lowered respect, I can't resist saying, bulls have it made in the shade.

However, exclusive rights are not guaranteed. Two or three bulls may be active among such large groups.

FIGURE 10.3. Large herd in the shade of a big tree.

Territories established at a waterhole also bring the owners into contact with large crowds. But stopping them from coming and going is usually difficult and involves a lot of chasing and calling. Additional energy is expended in chasing after bulls navigating the territorial gauntlet. Here again, territories with shade trees, which are common near water, can attract many animals, which often wait a considerable time for their turn to drink. As wildebeest tend to go to water after the morning feeding peak and in the middle of the day, many will tarry in the shade both coming and going.

Another favorable time for rounding up a herd is early evening after the afternoon feeding peak, as the gnus are preparing to rest and ruminate for several hours after nightfall. But the transition also marks a spike in chasing and calling and fighting, culminating in the Big Hum during the rut (to hear it: www.rarespecies.org/BigHum.mp3). The activity dies down gradually, reaching a low point when most animals are lying and ruminating. It builds up again when the resting animals resume grazing but rarely reaches a crescendo on dark nights. It's another matter on moonlit nights: rutting activity is just as intense as during daylight. On dark nights, too, a general movement in one sector will provoke an outbreak of calling and herding behavior marking the spot. The quietest time, before daybreak, often ends abruptly as the resting herds rise and begin grazing soon after dawn. Bulls with herds try to keep them bunched but usually fail. Groups spread out and graze or may aggregate and move in search of better grazing. The commotion created by territorial bulls builds to another climax marked by the Big Hum.

As there are bulls spaced out on territories ahead of moving armies, so are there bulls who stay put after all the wildebeest have passed. If an aggregation stayed long enough to reduce the pasture to a short sward, you would think it was high time these diehards moved on. But wait. Come evening, if the aggregation has been feeding in tall grass, wildebeest in search of the safest place to spend the night may come back to the mown pastures. This is another tactic with a potential payoff.

How long a bull holds a particular territory varies from less than fifteen minutes to days, even weeks. On open plains with few distinct landmarks, nearest neighbors define territorial boundaries. Interactions with his abutters limit how far a territorial bull can go in any given direction. David Western (pers. comm. 1986) tells of a single bull in Amboseli National Park that accompanied a small herd of females and young throughout their home range of 11 sq. km over a period of eight months. Challenge rituals maintain the status quo. As noted earlier, challenge rituals usually take place well inside a territory rather than on the boundary, as in most associated antelopes. Territorial wildebeest do not define their boundaries with dung deposits and preorbital gland marks. The central stamp is the olfactory hub of the property (see fig. 8.4).

The take-away point is that, absent topographic restraints, a bull's territory extends until he encounters an equal and opposite force in the form of neighbors defending their space. The greater the density of females and young, the more territories are compressed and the closer the spacing between bulls. It can be as little as 30 m in a dense concentration, or even less when two bulls operate in the shade of a big tree.

The willingness of bulls to stay alone for extended periods in areas of high grass that could conceal a lion demonstrates that posterity is of greater concern than longevity. In other words, the drive to reproduce trumps concern for safety. At the same time, the presence of neighbors also reduces individual risk and should make each feel less vulnerable.

MALE REPRODUCTIVE PHYSIOLOGY AND BEHAVIOR DURING THE RUT

Watson[1,2] did not find any seasonal change in the testis size and sperm production (spermatogenesis) in a shot sample of forty-eight mature bulls in the migratory Serengeti population, nor did he find any change with increasing age. However, he added the caveat that further information was required to confirm the absence of seasonal variation.

Observations during rutting peaks strongly suggest that heightened testosterone and sperm production underlie the greatly increased sexual activity of territorial bulls. Not only are changes in testes size found in other seasonally breeding antelopes, including springbok and impala, and related black wildebeest[3] and blesbok,[4] but Attwell[5] found greatest testis development at the beginning of the rut in his study of blue wildebeest reproduction in Natal. Maximum testis mass coincided with the rut (April), after which it decreased by 48.6 percent to a minimum value in July.

Here are some behaviors that are common during the rut and much less common at other times. They show how testosterone-driven bulls often become.

Bulls advancing to meet incoming cows often have a partial erection.

The presence of cows in estrus often stimulates nearby bulls to increase, even double the tempo of their calling, and they may foam at the mouth in their excitement. Copulations by such oversexed bulls leave saliva marks on the shoulders of mounted cows, which serve to identify cows in estrus to human observers.

Attempts to mount cows on the run, never successful, are broadcast by what I dubbed the "passion call," which sounds quite like a drawn-out, modulated belch (see fig. 10.13).

Bulls, including territorial, yearling, and subadult individuals, will sometimes attempt to mount calves, mostly ones that have lost their mothers. The stimulus that prompts this is obscure. Possibly a substitute object easier to mount than unreceptive cows? I have even witnessed rapes. Calves, too, mount one another and sniff the backsides of estrous females. Yet youngsters under two years old have not reached puberty.

Bulls thwarted in attempts to mount a cow in estrus sometimes rise on their hind legs with full erection in a spectacular display that I dubbed "flashing" because it reminded me of cartoons of men in raincoats displaying to female passersby (fig. 10.4).

Spontaneous ejaculations, in which a bull suddenly humps his hindquarters and makes the thrust that completes copulation but without rearing.

FIGURE 10.4. A frustrated bull "flashing" before an uncooperative cow.

Herding and chasing cows, fighting, and the crescendo of calling that produces the Big Hum are all at their peak during the three-week rut and far less pronounced the rest of the year. The close herding of females and young that is a hallmark of the rut was mentioned. Combat is much more common than challenge rituals wherever females are present. The general hubbub of contending bulls is punctuated by the loud knock of horns as trespassers are stopped in their tracks by herd defenders. Being willing and able to bash heads is a rigorous test of fitness, for the impact of two animals each weighing 200 kg (450 lb.) colliding on the run can break horns. And not just the narrower outer section but even the massive boss. In fact, on rare occasions the impact breaks off a whole horn together with the underlying skull, leaving a bull with a hole in his head (fig. 10.5).

I couldn't believe it until I saw it and photographed it for myself during the rut of 2009. Since then I've seen or heard of a half dozen instances. The chances of a wildebeest surviving with a hole exposing its brain must be about zip. I mentioned earlier the resilience of a wildebeest's horns, which is so great that blows with the back of an ax on a fresh skull bounce off harmlessly. So the torque or "bending moment" great enough to break a horn must be tons per square inch. I've made a habit of recording broken horns while taking sex and age samples

FIGURE 10.5. Bull with a hole in his head.

but primarily when recording the ratios of females and young while ignoring bulls with intact horns. So I can only report the incidence of broken horns among females from a few samples. It is certainly rare, since cows rarely butt heads, but obviously it happens—for example, while trying to protect offspring from rambunctious bulls. An idea of just how rare is illustrated by a random pick of my sex and age data. Only one of 492 cows had a broken horn, equal to 0.002 percent.

Bulls with broken horns are at least two or three times more common. But they can still engage in the rut if both horn bosses are intact. The knobs absorb the impact of a charge. I have always maintained that bulls missing one complete horn could only hold space in a territorial network if they avoided butting heads. During the 2011 rut, I witnessed and photographed an exception to that rule, too. A territorial bull in his prime but missing his left horn was just as active as his neighbors, to the point of bashing heads without giving any ground (fig. 10.6). As far as I could tell, it didn't faze him.

The rut also stimulates other old and poorly armed bulls to try their luck. The example of the old bull branded E who emerged from retirement and continued to hold territory for a year was noted earlier (see fig. 8.3). Here again, one suspects a surge in testosterone production as the trigger. But longitudinal fecal-hormone sampling of known bulls, comparable to the study of female reproductive cycles reported below, are needed to monitor endocrine levels before, during, and after the rut. Our plans to carry out this research on radio-collared bulls in the

FIGURE 10.6. A brave bull that refused to be sidelined by a missing horn and even dared to knock heads with territorial neighbors.

resident Kirawira population had to be given up for logistical and political reasons.

CONTRASTS IN THE BEHAVIOR OF HAVES AND HAVE-NOTS

A bull with cows on his territory and nearby rivals is fully occupied with holding onto them and acquiring more. He doesn't graze or lie down but keeps moving around and calling. He reacts immediately to incursions by his neighbors, cantering to meet them with tail waving and banging heads with any that fail to give way. When a nearby herd is under siege by invading neighbors, the bull will round up detached cows and bring them back to his group. But the bigger the herd, the harder it becomes to keep all the members together, as neighbors besiege his string from all sides, making it impossible to keep them all at bay. The average number in the rutting clusters photographed from the air, about thirty head[1] may be a sort of cutoff point. But there are so many variables that giving hard and fast numbers is pointless.

So our focal bull finally loses all his cows and finds himself alone on his patch. Likewise, his nearest neighbors have seen their herds depart as the rank and file took it into their heads to move on. In very short order the bulls go into maintenance mode, figuratively shifting into low gear. Our bull begins grazing and presently begins chewing his cud either while standing or lying down. But he is still mindful of his territorial status and the need to maintain his place in the network, which entails all the associated behaviors and displays described in chapter 8. He engages in challenge rituals with his neighbors. A noticeable difference from interactions at other seasons is that combat is more frequent, indicating a higher level of aggression fueled by testosterone.

By resting and feeding in this way when no cows are in range, wildebeest bulls are able to stay fit and continue competing for mating opportunities. This is very different from most seasonally breeding ungulates that have been studied, such as bison, goats, sheep, and deer, which compete to exhaustion[6]. But those species all live in the north temperate zone and have male dominance-hierarchy mating systems, which as noted in chapter 2 are more competitive than territorial systems.

Moreover, territorial wildebeest that have competed to exhaustion in a major rutting aggregation may sojourn for a few days in a bachelor herd before reentering the territorial ranks. Such was the case with one of the four radio-collared bulls I followed for up to five days during the 1979 and 1980 ruts. This bull spent alternate days in bachelor herds, apparently exhausted by a full day on territory in the midst of the migration.

In the wake of a migrating army, there are usually a few stragglers on the battlefield, mainly bulls with disabling injuries sustained during the rutting melee. Broken or dislocated forelegs are most common. Often these limping wounded join together in small bachelor herds of two or three or more bulls.

The surest sign that the rut is over is the presence of resting herds containing more than one adult male. It means the territorial network is no longer intact. In 2011, the peak rut among many thousands of migrants that passed through the Grumeti Reserves began at the end of May and was winding down by mid-June. That is to say, the full moon of June 15 marked the end of the peak rather than the beginning, as Sinclair's lunar trigger hypothesis[7] predicts.

Sinclair's 1997 paper in *Nature* proposes that the moon's influence on the timing of the wildebeest's rut is "the first case of a strong relationship between conception and the lunar cycle in mammals" (832). Yet the dif-

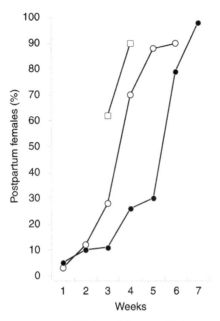

FIGURE 10.7. Timing of the 1973 birth peak among small herds in three different areas of Ngorongoro.

ferences in the timing of the rut just within the Serengeti region, already noted, are prima facie evidence that the full moon or any other phase is not the Zeitgeber. The rut in Ngorongoro and in the Serengeti is more often out of than in sync and not tied to intervals of 29.5 days. Even in our 1973 Ngorongoro calving observations, birth peaks differed by up to a month in different parts of the Crater, from early January to late February (fig. 10.7).[8] Birth peaks on different calving ground concentrations also varied but by fewer days. (See chapter 11.)

To estimate the timing of the preceding rut and the lunar phase, I subtracted 242 days, now the accepted (eight-month) gestation, making May 7, 1972, the peak of the rut prior to the January 21 calving peak, and June 10 the peak rut preceding the February 15–23 calving peak. These dates are about midway between the two full moons of April 28 and May 28 and May 28 and June 26. So no full moon influence is identifiable.

In his table 2, Sinclair[7] calculates the timing of the preceding rut from records of first calves recorded in the Serengeti by Warden Myles Turner from 1956 to 1965. Subtracting 255 days (an assumed 8.5-month gestation) from the calf sightings, the conception dates range from March

1 to April 17. As the rut is consistently three to four weeks in May–June or June–July, these dates simply reflect the fact that early births often occur months before calving peaks, which generally occur in February give or take a few weeks.

REPRODUCTIVE PHYSIOLOGY OF SERENGETI FEMALES

After experiencing the rut in Ngorongoro Crater and in Serengeti NP in 1963 and 1964 while pursuing my dissertation research, I became convinced that the calling of the bulls played a part in synchronizing the mating peak. The chorus of the Big Hum reminded me of thousands of croaking frogs; as the calling of frogs was known to stimulate females to produce eggs and mate, one could reasonably assume that the croaking of wildebeest bulls could have a similar effect. It was only many years later, after the turn of the century, that the opportunity to test this hypothesis came my way, in the form of a grant by the Smithsonian Institution Fund for Scholarly Research for a study of the reproductive physiology of the Serengeti wildebeest population, "The Causation of Reproductive Synchrony in the Wildebeest *Connochaetes taurinus*"; principal investigators Steven L. Monfort, Richard D. Estes, and Katarina V. Thompson.

The site chosen for the research was the Grumeti Game Reserve, a small (5,000 sq. km) but important part of the Serengeti ecosystem abutting the Western Corridor of Serengeti National Park (see map 7.1). As noted in the introduction, we set up camp and built a fenced enclosure at the foot of Sasakwa Hill to hold the wildebeest cows captured for the study. Preparations for carrying out the research, beginning with buying a used Land Rover in November 2001, were drawn out and subject to frequent delays. Fencing the 25 ha boma had to be done from scratch due to the unavailability of game-proof fencing in Tanzania. Ten strands of high-tensile wire were strung along sisal poles implanted four feet apart. To make the fence look solid, strips of split sisal poles were used to fill the gaps (fig. 10.9).

Allison Moss, a graduate student of Monfort's, signed on to carry out the fieldwork for her doctoral dissertation at George Mason University.

By the time everything was ready to begin the capture operation, the wildebeest migration had already passed through the reserves en route to Kenya. It was necessary to wait for their return, which finally came in November. Monfort came from the United States to participate in the capture of wildebeest cows. A total of 18 were anesthetized and

FIGURE 10.8. Steve Monfort and Allison Moss with the Serengeti NP veterinarians who darted the wildebeest captured for the study of their reproductive physiology.

FIGURE 10.9. Captive cows and calves in the main enclosure.

transported successfully, using darts containing M-99 (etorphine hydro-chloride). The sample included 15 pregnant cows and 3 unbred yearling females.

The boma enclosed a patch of the natural savanna habitat. But 25 ha of grass were insufficient for the captives, especially after the 15 cows had calves in February 2003. So additional grass was cut and provided daily, supplemented with improvised livestock feed. Each animal was individually marked using colored ear tags.

While my focus in the study was the role of calling in synchronizing estrus cycles, this proved to be just one of the unknown aspects of wildebeest reproduction the study addressed. Though short rutting and calving peaks are defining wildebeest traits, whether wildebeest cows are strictly seasonal breeders was unknown. After all, some 20 percent of calves are born over a five-month period following the calving peak. Did cows ovulate just once (monoestrus) or, if unbred, repeatedly (polyestrus)? Was ovulation induced by external influences (e.g., bull calls), or was it spontaneous? How long did lactation inhibit estrus following calving? Even the duration of gestation was not precisely known: estimates ranged from eight to nine months.[2,9,10]

Among all species in the tribe Alcelaphini, such basic reproductive parameters are known only for the blesbok *(Damaliscus dorcas phillipsi)*.Captive blesbok females were found to be seasonally polyestrous, with a mean gestation of 240 days, but other parameters of the estrous cycle were not quantified.[11]

Assessing hormone breakdown products excreted in the dung is a noninvasive method that allows the study of reproductive patterns, such as seasonality, pregnancy, and ovarian cycling. Monfort was a leader in developing the procedure.[12] The research protocol required the droppings of each cow to be collected at least three times a week. That required observing the chosen individual until she dropped a cluster of pellets, finding her deposit in the grass, collecting a small sample, and putting it into a little plastic bag labeled with her name and the date. Several men from nearby villages were employed for the task and soon became expert at recognizing each cow (initially by ear tags), noticing when she defecated (not obvious in wildebeest, which just raise the tail slightly and dribble their droppings), and making sure the pellets they collected were not another cow's. This task became easier within a week or so of capture, as the cows quickly lost their fear of people and hardly reacted to their near-presence in the enclosure (see fig. 10.9). Plus their grazing reduced the height of the pasture.

Analysis of the dung samples had to be carried out in Monfort's lab at the Smithsonian Conservation and Research Center in Front Royal, Virginia. But preliminary processing was carried out on-site. Dung samples that are simply dried and stored in little plastic bags can be used for hormonal assay; Simon Mduma employed this method to detect pregnancy from progestin levels during his wildebeest dissertation research.[13] He collected fecal samples from pregnant females culled during months 4 to 8 of gestation and from nonpregnant animals accompanied by newborn calves following the birth peak.

A major advantage of fecal monitoring of the same animals is the ability to monitor individual endocrine patterns for an extended period. The far more elaborate and labor-intensive protocol for the scat samples of our captive cows began with keeping the samples on ice until they could be transferred to a propane-operated freezer for storage until extraction. The extractions were conducted in a small tent that doubled as a field laboratory, with power supplied by a portable generator. After the samples were thawed, a tiny amount (0.25 g of feces) of each one was dissolved with 5 ml 100% ethanol in a test tube. Then the batch of samples was boiled for thirty minutes, followed by spinning for half an hour at 2,500 rpm in a centrifuge. What was left after all this was poured off into clean glass test tubes and dried completely under compressed air. Finally, 1 ml of ethanol was added to each tube, which was then vortexed for sixty seconds. But wait, that's not the end of it. The last step was pipetting 850 μl of the extracts into labeled 12 × 75 mm plastic test tubes, to be dried completely, capped, and stored at room temperature for up to six months, or until analysis in Monfort's lab.[13]

The temperature in the tent in which Allison carried out this elaborate protocol rose to the mid-90s (F) in the midday hours, and I remember counting dozens of horseflies (preferable to tsetse flies, mind you) resting on the inside canvas a few feet from where Allison was working. Fortunately, they were largely inactive; otherwise, Allison might have needed a transfusion!

All the adult females were pregnant and dropped calves the February following capture. The entire herd was left in the 25 ha enclosure until the end of March 2003, when the youngest calf was two weeks old. Then they were all anesthetized and transported to one of three new enclosures, each 100 × 100 m. Calves remained with their mothers. The adults and the three subadults were randomly distributed in three groups of five each. At the beginning of May, a territorial bull from the Kirawira resident population was captured and introduced to one of the groups.

The three bomas were spaced roughly a kilometer apart to keep each group isolated and effectively screened from one another by patches of riverine forest. It was particularly important to keep the control group (1) from seeing, hearing, or smelling males. Beginning on May 3, about a month before the onset of the Serengeti rut, Group 2 and Group 3 (+ bull) were exposed continuously for three weeks to the bull rutting calls, transmitted via CD recordings over battery-powered game-caller speakers. To keep the captives from becoming habituated to the calls, the different sound tracks were repeated in random order and played in different locations.

I had tape-recorded the rutting calls while observing ruts dating as far back as 1965. At peak calling times, with thousands of bulls producing their basso grunts and honks, the air reverberated and pulsated, against a background roar that reminded me of heavy surf breaking against a rocky headland. It was the sustained rumble and roar that I labeled the Big Hum. The calls of nearby bulls, often individually distinct, punctuated the background noise. (Sample recording at www.rarespecies.org/Individual.mp3). Sitting in my Land Rover surrounded by thousands of animals, I sometimes felt I was drowning in sound.

The effect of the rutting calls on the two groups of cows was largely invisible to the human eye, despite behavioral observations of one-half hour to one hour duration that were routinely carried out at random times of day throughout the study. This research was intended to determine whether the captive herds of females exhibited a cohesive social structure with clearly defined dominance hierarchies (confirmed in all three groups).

For three weeks before, during, and after females were exposed to the rutting recordings, dung collections were increased to daily. Matching behavior with estrogen and progestin levels had to wait for the fecal samples to be processed at Front Royal. Before behavioral observations could be correlated with ovulation, basic endocrine profiles had to be established for each cow. (That finally happened two years after the fieldwork ended. Wooed by the head chef of the newly constructed luxury Sasakwa Lodge atop Sasakwa Hill, Andy Clay, Allison married him, became pregnant, and then couldn't work in the Front Royal lab for fear of radiation effects on the fetus.)

Results of the behavioral study reported in her dissertation[14,15] have yet to be published. One would expect to see behavior most obviously related to estrus in Group 2 + bull. But the bull showed so little interest in the rut that he didn't even join the recorded chorus. He was more of

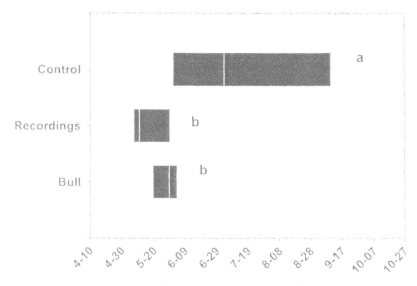

FIGURE 10.10. Synchrony of the first estrous cycle in female wildebeest held within their natural range in three treatment groups.

a dud bull than a stud bull. As he came from the resident Kirawira population, which breeds a month earlier, maybe he had already shot his bolt, as the saying goes. Nevertheless, he was seen to cover cows on several occasions, and all of those in his enclosure became pregnant. One cow was observed mating once, another on two consecutive days, and a third cow was observed mating three times, the last two on consecutive days. A fourth cow was observed to copulate on June 5 but failed to conceive, as the bull mounted her again on July 15 and 16. She calved the following March, just after all the captives were released.

When the laborious process of analyzing the fecal hormones was finally completed, the results showed clearly that ovulation within the two groups exposed to the rutting calls was synchronized much more closely than in the control group (fig. 10.10).

During the three weeks immediately following cessation of the recordings, progestin concentrations in the control group's excretions were significantly lower compared to both the Recordings Only and the Recordings + Bull groups, indicating that the recordings stimulated production of reproductive hormones.

Fecal progestin concentrations during the 34 weeks following the 3-week treatment interval continued to be lower in the control group compared to the Recordings Only group. (As the cows in the Recordings + Bull group were all pregnant, they were excluded from this analysis.) But the average number of estrous cycles was also greater in the Recordings Only group compared to the Control group: 9.2 cycles compared to 4.0 cycles. Moreover, the average date of the last estrous cycle in the Recordings Only group was January 11, compared to October 26 in the control group.

Summary of Results

1. Wildebeest are spontaneous ovulators.
2. They are seasonally polyestrous.
3. The average duration of estrous cycles was approximately 23 days.
4. For nonpregnant females, estrous cycling was observed for an average of 203.5 ± 29.9 days (average date of cessation, December 3).
5. The average duration of gestation, approximately 242 days, was slightly shorter than previously published estimates (9 months; 8–8.5 months).[9,10]
6. Of the 15 cows that were pregnant when captured, all resumed estrous cycling within 3.5 months (from 85 to 120 days, average of 102.4 days) after parturition. The resumption of ovarian activity coincided with the breeding season previously reported.
7. All unmated cows with surviving calves (n = 5) resumed estrous cycling while still suckling their calves.
8. Of the mated cows with calves in Group 3 (n = 3), all discontinued suckling within six days after introduction of the bull.

The Smithsonian study provided persuasive evidence of the important role that male rutting calls play in modulating estrus cycling in the Serengeti wildebeest.[14,15]

Synchrony among females, the robustness of ovarian-endocrine response, the frequency of ovulations—and hence mating opportunities—and the duration of the potential breeding season were all positively affected by exposure to the calling of the bulls. Cows exposed to vocalizations first ovulated within four weeks of exposure, and their ovulations were clustered within an approximately three-week interval. This distribution profile would translate to the synchronized birthing pattern typically observed in migratory wildebeest.

In contrast, the control females exhibited more sporadic, unsynchronized estrous cycles. Had each cow conceived with her first ovarian cycle, births would have been spread across a twenty-three-week interval. That would certainly be inconsistent with producing a synchronized two- to three-week calving peak, and a peak birth season lasting over three months would surely increase calf vulnerability to predation.

The three-week exposure to recordings alone elicited more frequent estrous cycles, and over a more extended time frame compared to the control group. This possibly ancestral breeding strategy would ensure that cows that failed to conceive during the rut become pregnant during the following months—even up to November—thus accounting for the 10–20 percent of conceptions that occur following the rut. Clearly, wildebeest are seasonally polyestrous. Moreover, this capability can be adaptive by enabling a population to change its breeding season in response to natural disasters or climate change. Thus in 1960–61 an extended drought caused near-total mortality of the eastern and western white-bearded wildebeest calf crops. The abundant rains in 1962 caused cows, now in good condition and freed of lactation anestrus, to breed two months earlier than usual, calving in December 1962 instead of February 1963. Over the next three years the schedule gradually returned to normal.[8,16] Rebuilding of the populations was promoted by an 80 percent yearling conception rate, compared to a normal 20 percent. This highlights the role of nutrition in reproductive physiology. Unbred females in good physical condition can begin ovulating earlier than ones in fair to poor condition and are also likely to produce more viable offspring.

Unexpectedly, all the captive cows ovulated by May 5, three weeks before the rut of the migratory population observed by Moss Clay.[14] Of the eleven cows that exhibited early estrous cycling, nine experienced their early cycle within nine days of one another (between April 22 and May 1). As these animals were distributed among the three geographically separated treatment groups, a shared social cue seems an unlikely trigger. However, differences in nutrition among migratory and our experimental herds could in part explain the early onset of estrous cycling. Supplementing the captives' diet with cut native grasses and livestock feed may have raised their condition above a "nutritional threshold," which accelerated estrous cycling relative to the migratory hordes. In addition, the study site was located in a particularly wet portion of the wildebeest's range. High pasture quality in this region may in part explain why the resident Kirawira wildebeest population breeds one to two months earlier than the migratory population.[10,17]

WHY SOUND TRUMPS SMELL IN SYNCHRONIZING
WILDEBEEST ESTRUS

In many other ungulates, male presence has been demonstrated to be an important modulator of female reproductive function, advancing the timing of estrus and shortening the anestrous interval[18-20]. In many species, olfactory cues have been implicated.[21,22] But the degree of synchrony within both captive wildebeest groups exposed to vocalizations did not differ from one another: the presence of a single bull did not increase ovarian function beyond the effects of recorded vocalizations alone. In wildebeest, with close to 500,000 females in the migratory population, for any cue to be effective in synchronizing estrous it must be simultaneously accessible to most of the females. Chemical cues face two limitations: first, the size of the Serengeti-Mara population would hinder the diffusion of chemical cues; and second, only male wildebeest test female urine by performing flehmen.[10] That means females do not monitor one another's estrogen levels with the vomeronasal organ. The mere presence of males is also an unlikely synchronizing cue, as females are exposed to males throughout the year. While male competition is heightened during the rut, the sound of thousands of bulls calling at fever pitch, the Big Hum, is audible over a wide area. The results of the playback experiments suggest it is mainly the sound that serves to stimulate and synchronize estrus. An intriguing possibility that remains untested is that the spontaneous ovulation of cows ahead of the rut stimulates gonadal activity in the bulls, whose increased territorial activity and calling makes for an effective feedback loop.

Discontinuance of suckling following introduction of the bull to the Recordings + Bull group is inconsistent with observations that calves continue suckling during the rut. Moreover, in a shot sample of hundreds of Serengeti cows, Watson[1] found that the mammary gland, which reached a maximum weight of 1,400–1,600 g from the end of January to May, continued to lactate while the udder shrank in the dry season to about 500 g in October. Data from studies of the blue wildebeest in Natal and Namibia are in close agreement.[5,23] So calves are only finally weaned at nine months (more in chapter 11).

The study of reproduction in the Serengeti population carried out by Watson[1,2] makes a stark contrast with the longitudinal study of fifteen captive cows just described. One hundred fifty-three cows were shot during the second half of 1964, throughout 1965, and during the first few months of 1966, so that Watson could collect their ovaries, measure fetal growth rates, gestation, lactation, postpartum anestrus, and so on. (If such ad libitum sampling now seems excessive, it was normal practice at the time for veterinarians, wildlife researchers, and managers to collect as many specimens as desired; it was also accepted parks practice to shoot game for staff rations in areas outside the park [Klingel pers. comm., 2013].)

The presence of an unfertilized and regressing corpus luteum (literally, "yellow body," formed immediately after ovulation that secretes estrogen and progestin), together with an active follicle in three cows shot while mating, gave Watson[1] the idea that cows might undergo a "silent heat" before true estrus. A regressing corpus luteum together with an active corpus luteum of pregnancy or of cycle were found in ten of fourteen cows collected between June 11 and July 7, 1965, supporting Watson's hypothesis that true estrus required the presence of a waning corpus luteum. These findings also support the evidence, already noted, of spontaneous ovulation prior to the rut in the fecal hormones of our captive cows.

MATING BEHAVIOR: FEMALES IN ESTRUS

The typically passive female sexual role is well exemplified in the wildebeest (fig. 10.11). The general impression given by cows in estrus is that they are rather bewildered by it all and desire nothing more than to remain inconspicuous and unharassed. Aside from occasional mild displays of coquettishness, as when a cow walks up to a bull to touch noses, lowers and shakes her head, then cavorts away as though inviting pursuit, the only sure sign of a female in season is the behavior of bulls and other wildebeest toward her. While the bull's interest is, naturally, the more obvious, yet other members of the herd also manifest considerable interest that seems to combine both sex and aggression. Cows hook at her and jostle her. They may mount her and each other. Yearlings and even calves follow behind, sniffing her, and may even attempt to mount. Notably absent in this general display of sexuality is any apparent inclination by estrous females to mount other members of the herd.

This same passivity distinguishes the actual act of mating. Instead of standing still, a cow was more likely to move out from under the mounting bull, resulting in a high percentage of apparently incomplete copulations. When closely pressed, she could gain a temporary respite by lying down (fig. 10.12). Such a performance might be expected from yearlings in estrus, which appear and behave more like teenagers than

FIGURE 10.11. Copulation—a successful mount.

FIGURE 10.12. A cow responds to sexually harassing bachelor males by lying flat out in the posture of a hiding calf. This illustrates the need for the order imposed by a territorial system.

adults, but it was a little surprising to observe that old cows were only slightly more responsive toward the bulls.

The above account was taken from a report of the first rut I observed, in 1963, which was submitted in a quarterly report to the National Geographic Society Committee for Research and Exploration. The rut I observed a year later, in the Serengeti, made the Ngorongoro rut look like a tryout for the main event. Instead of some twelve thousand Crater wildebeest, hundreds of thousands were involved in the Serengeti rut. The fact that it was happening while the population was migrating exacerbated the turmoil. It was general movements that revved the bulls' calling to the fever pitch of the Big Hum. The noise of the Crater rut was a faint echo in comparison. During the height of the Ngorongoro rut, I estimated the density of territorial males in the thickest concentrations at three per hectare. During the 1998 Serengeti rut, I tried to count the territorial males in a 100 m quadrat marked by long striped poles I set up in the midst of dense aggregations. From atop my vehicle, I counted up to ten active bulls in a quadrat. A density of ten territorial males per hectare is comparable to the spacing of males in the three lekking antelopes: kob 15–35 m,[24] lechwe 15 m,[25] topi 25–40 m.[26]

Some researchers have suggested that migrating wildebeest may be demonstrating a form of lekking.[27-29] But none of the four criteria that define leks—no male parental care, males clustered on an arena in territories that offer no resources, to which females in estrus come to mate with the fittest males—aptly describes the wildebeest system. No male parental care applies generally to bovids with polygynous mating systems; females can exercise mate choice in other polygynous systems; and wildebeest territories in migrating aggregations include patches of the preferred grazing. The leks of topi and kob, which include 30 to 40 up to 100 males,[30] are established in traditional hotspots where large aggregations gather. There is no evidence of such an arrangement in wildebeest. Where migrants concentrate is unpredictable, and the establishment of a territorial network is spontaneous and ephemeral. One known advantage of lek systems is that estrous females are less subject to harassment by gathering on leks than on conventional territories. Wildebeest cows minimize harassment by relaying between the groups held by territorial bulls.

Wildebeest cows mate many times while in estrus. Exactly how long estrus lasts, as judged by willingness to stand for copulation, has yet to be determined but is probably at most twenty-four hours. If undisturbed, a cow will choose to mate repeatedly with the same bull. Here are representative examples I recorded while observing rutting behavior at different times and places.

6.21.99. At 14:42 a bull mounts an estrous cow twice (++) and again 8 sec. later (+). The cow runs to the next herd, and that bull tries to mount her on the run (-). She runs on. At 14:45 a third bull tries to mount (-), but she runs on again.

6.21.99. At 10:25 a territorial bull tries to mount a lying cow, from the front instead of the rear. She gets up and runs. A white mark on her shoulder shows she has previously copulated.

6.30.99. On the Sabora Plain, covered by a sea of wildebeest, a bull mounted a cow five times in one minute at 06.38. All mounts looked to be complete with ejaculation as the bull made the thrust that straightens the S-bend (sigmoid flexure) and fully extends the penis. The cow reacted by hunching her hindquarters with tail out, a typical postcoital behavior.

6.2.03. In a herd containing 12 cows, the herd bull mounts a cow at 17:53 (complete = +), and again 18 sec. later (+). A different female approaches the bull, rubs her horns on his rump and gently butts him. The bull runs after her and the first cow follows him. At 2.03 min., he mounts the first cow (+), again at 2.50 (incomplete -), at 3.13 (+), 3.57 (+), and 4.15 (+). At 5.21 min. the herd starts leaving. The bull butts heads with his nearest neighbor. The estrous cow stays in his remaining herd of 5 females. No further mating observed. As this example shows, a group may include a second (rarely even a third female), with whom "Lucky Pierre" takes turns copulating (I didn't see it happen in this case). It is possible that the presence of one estrous cow stimulates ovulation in another.

To my knowledge, the multiple copulations that distinguish the wildebeest's tribe from most other antelopes (cf. accounts of Antilopini, Hippotragini, Tragelaphini),[30] remain unexplained. The observed preference of a cow in estrus to stay and keep mating with the same bull is an example of female choice. But the chosen mate is often besieged by his nearest neighbors to the point that he can barely find time to copulate. In a mating episode I filmed some years ago, the bull is so distracted by his neighbors' end runs that his efforts to reconnect with the estrous cow, who is patiently waiting for him, are thwarted. Four tries to rejoin his herd of eight cows are interrupted in order to bash one neighbor who keeps moving in behind him. When he finally mounts the

FIGURE 10.13. Estrous cow escapes attempted mount, provoking the bull to voice the "passion call."

cow, he's in such a hurry that only one of four tries is complete. Often enough, horning in by the neighbors succeeds in dispersing a defended group, and the estrous cow ends up in a different herd. Surprisingly often, bulls trying to add and avoid losing cows fail to detect her before she moves on. But I have also seen a cow relaying from herd to herd copulate with up to five different bulls. Estrous cows can avoid copulations by moving out from under a mounting bull or simply running away, provoking the passion call already described (fig. 10.13). Some such cases of avoidance could indicate the cow was no longer in estrus.

REFERENCES

1. Watson 1969.
2. Watson 1967.
3. Skinner, van Zyl, and van Heerden 1973.
4. Skinner 1971.
5. Attwell 1982.
6. Putman 1996a.

7. Sinclair 1977b.
8. Estes 1976.
9. Talbot and Talbot 1963b.
10. Estes 1991a.
11. Marais and Skinner 1993.
12. Monfort 2003.
13. Mduma 1996.
14. Moss Clay 2007.
15. Moss Clay et al. 2010.
16. Estes and Estes 1979.
17. Kingdon 1982.
18. Knight and Lynch 1980.
19. Iason and Guinness 1985.
20. McComb 1987.
21. Sempéré et al. 1996.
22. Signoret 1990.
23. Berry 1997.
24. Leuthold 1966.
25. Schuster 1976.
26. Monfort-Braham 1975.
27. Balmford et al. 1993.
28. Hopcraft et al. 2010.
29. Clutton-Brock, Deutsch, and Nefdt 1993.
30. Gosling 1986.

CHAPTER 11

The Calving Season

Birth and Survival in Small Herds and on Calving Grounds

Eight months after the rut, the cows that became pregnant during the three-week peak produce their calves. Assuming a 95 percent pregnancy rate and that 80 percent of the cows bred during the rut, over 400,000 calves would be expected in a population of 1.3 million wildebeest. Given a carnivore population of about 3,000 spotted hyenas and 2,000 lions, plus a few hundred cheetahs, leopards, and wild dogs—the only predators that actively hunt wildebeest calves—it is obvious that they can only consume a small percentage of the calf crop. Glutting their predators is rightly considered the primary advantage of having as restricted a breeding season in the tropics as ungulates in the North Temperate Zone with a short growing season. This extraordinary shortening of the birth season has evolved because wildebeest calves do not hide, yet retain the tan color inherited from hider ancestors that is conspicuous against the gnu's dark uniform.

To understand how the two wildebeests came to be the only antelopes to entirely abandon the ancestral hider-calf strategy, first the waypoints in the transition from hider to follower calves have to be considered. As discussed in chapter 3, the fact that hartebeest calves depend on concealment like all other antelopes indicates that the transformation to a follower system happened after the Alcelaphini became a distinct tribe in the late Pliocene a few million years ago. Stages in the transition can be seen in the topi's genus *(Damaliscus)*, in which the blesbok has follower calves but retains elements of the hider system, notably calves that are somewhat slower to become mobile and do not follow as closely as wildebeest calves (see chapter 3). It is interesting to note that all alcelaphine calves retain the ancestral tan color and, in fact, look remarkably alike considering the pronounced contrast between adults.

The wildebeest's habit of forming large, mobile aggregations favors shortening the hider stage, as mothers guarding hidden young can be left behind when an aggregation changes location. The difficulty is compounded if the population is migratory. The herding instinct is so strong that I have known wildebeest to abandon calves that were unable to accompany them when their aggregation moved away. The absence of cover for calf concealment would be another major problem for wildebeest. Mothers of hider species seek isolation and need cover to conceal newborns. Migratory wildebeest congregate on the shortest, greenest pastures during the calving season, where the locations of hiding tan calves would be pinpointed. Natural selection favoring survival of the most precocious calves led to an ever shorter hiding stage, culminating in offspring capable of following their mothers within an hour of birth. The eight-month gestation, adapted to peak growing seasons at both ends, meant that most calving would occur while the migration was aggregated on short green pastures. Further shortening of the birth season to the present three-week peak could have come about because survival of calves born during the peak would be higher than those born before or after the peak. As the following account of the calving season shows, the explanation is simple: the presence of calves in their most vulnerable newborn stage is obscured in nursery herds containing older calves; the more, the better. In other words, calves serve as substitutes for the vegetation that conceals hider calves.

From this viewpoint, the wildebeest's follower-calf system can be understood as a consequence of its adaptation to an ecological niche entailing migration between distant wet-season and dry-season pastures in arid and semiarid African savannas. As research has shown that survival rates in small herds are much lower (discussed below), resident populations should be considered offshoots of migratory populations rather than the other way around.

The association of wildebeest of the same sex, age, and reproductive status, described in chapter 8, is of greatest reproductive significance for the species in the gathering of pregnant females on calving grounds during the birth peak. Instead of seeking isolation like hider species, cows of follower species seek one another's company. The tendency of pregnant cows to associate and to remain in company during parturition

FIGURE 11.1. Ngorongoro calving ground.

leads to the establishment of calving grounds (fig. 11.1). These are not specific, traditional locations, although migratory wildebeest may return to calve in the same general area year after year. The Serengeti population returns to and calves on the short-grass plains every year unless the rains fail, but there are no fixed calving grounds. They happen on preferred feeding grounds, typically the shortest, greenest pastures available. In Ngorongoro, the wildebeest population usually concentrates on the floor during the rainy season, in areas with well-drained soil dominated by microperennial grasses. But whenever the tall grasslands of the Crater's slopes and hills are short, notably after wildfires during the preceding dry season, most of the wildebeest concentrate and calve there. Given a choice, wildebeest prefer the optimal view afforded by open slopes over flatland for calving. During the 1964 and 1965 seasons, many cows calved on the uppermost slopes of the burned uplands (fig. 11.2). Day after day, cows came to these grounds to calve, afterward withdrawing and joining nursery herds lower on the plateau.

Cows begin calving soon after dawn and continue until midday. In a sample of 129 Ngorongoro births known to have begun within the hour, only 17 were born after noon; 4 were born after 1400 h and none after 1600 h. Six were born at first light between 0500 and 0600. Why this morning peak? The later in the day a calf is born, the less time it has

FIGURE 11.2. *(Two-Page Spread)* Behavior on the calving grounds: a) Morning
maternity ward on upland above the Crater floor; b) Early stage of labor: emerging front
feet with yellow hoofs; c) Cow in advanced labor on a calving ground aggregation, front
legs out and head emerging; d) Calf dropping to ground; e) Newborn calf; f) Mother sniffs
and licks her newborn; g) Calf attempting to gain its feet; h) Failed attempt to stand;

FIGURE 11.2. *Continued* i) Still shaky but standing in under 10 minutes, heading for mother who exchanges grunting calls with offspring; j) Searching for the udder; k) Success: first suckling; l) Mother following newborn to keep it from attaching to any other nearby wildebeest; m) A mother fending off another cow's newborn; n) First-day calves cavorting; o) Proper following position next to mother; p) Calves begin experimental grazing within the first week.

to gain strength and coordination and become imprinted on its mother before hyenas, the main predators on calves, begin their nightly hunt. Surviving the first night is a major test of fitness.

The following account of the birth and survival of wildebeest calves is based largely on observations of the 1964 and 1965 calving seasons in Ngorongoro, as part of my dissertation research, and on a follow-on study of the 1973 season.[1,2] In 1964, the process of calving and calf predation were the focus of interest. In 1965 we monitored calving and survival rates in large aggregations during the birth peak.

> For the record, after getting married on 16 Dec. 1964, in a mission church near Arusha, and staying that night at the Lake Manyara Hotel, we spent the next month monitoring the calving season in the Crater by sampling 500 cows a day. We only left on our honeymoon in mid-January, to spend two months driving around southern Africa observing wildebeest populations while slowly wending our way back to Tanzania. One event that imprinted the memory of Runi's first days in the Crater was coming upon a hyena catching and performing a caesarian on a highly pregnant cow. Before our horrified gaze, a full-term calf dropped to the ground while the mother was still running. As it began struggling to get up, a chivalric impulse to shield my bride from the denouement led me to intervene but not before the hyena had chewed off both the calf's ears.

During the 1973 season, we focused on the rate and success of calving in small herds in comparison with aggregations. Observations of the 1979–81 Serengeti calving seasons included comparison with the Ngorongoro population (unpublished) while living at the Serengeti Research Center in Seronera, and from 1996 to 2002 I compared Ngorongoro and Serengeti offspring survival rates three times a year (Estes).[3] In between these times and during one- to two-month annual Serengeti stays from 2007 to the present, I sampled Serengeti offspring survival rates as opportunity afforded.

PARTURITION

Cows in the process of calving are easily spotted, in the first stage by the "bag of waters," or amnion, which acts as the opening wedge through the birth canal. It looks like a white balloon and may be as large as 14 cm in diameter. Usually it breaks by the time the feet have appeared, whose canary yellow hoofs are again conspicuous. The cow holds her tail out and arched or displaced to one side. During actual labor—that is, while contractions of the diaphragm and stomach muscles are combined with uterine contractions—the cow lies on her side, frequently rising to her knees or feet to change sides, which probably helps to bring the fetus into the correct delivery position. In going down, the mother occasionally lies right on a partially emerged calf, bending it into impossible angles, but apparently with no ill effects.

During the heaviest labor accompanying the passage of the head together with the forelegs, contractions come at intervals of one second or less and are powerful enough to lift the animal's hindquarters off the ground, while the pressure on the calf bows up the tail rostrum and completely everts the vulva. Her legs are held stiffly out from her body and raised, while the neck and cheek are pressed flat to the ground, although the usual position in the earlier stages and between contractions is with the neck and head turned back over the shoulder. Since adult wildebeest seldom lie flat on their sides at other times, the position during labor is noticeable.

Between bouts of labor, the cow behaves quite normally—though more alert and readily disturbed than other nearby animals. She may lie down, graze, or ruminate for another half hour or so after the calf's feet have appeared. Cows seldom lie flat for more than a minute at a time but roll back onto the brisket, rise to the knees, and resettle on the other side, or stand up. The appearance of more and more of the calf's legs without visible labor, noted on a few occasions, indicates that contractions of the uterus may continue unaccompanied by the abdominal muscles.

If need be, parturition can be interrupted at any stage for up to an hour, during which visible contractions cease. This is obviously adaptive to predation. But once the head and shoulders have emerged, a wildebeest cannot run fast enough to escape predators with a dangling calf swinging behind her. The IMAX film *Year of the Wildebeest* shows a lioness easily catching a wildebeest in this predicament.

The interval from passage of the head to final expulsion is relatively short, compared to the early stages. In 16 of 20 births, the calf was dropped within three minutes (± 2.3 min.), usually following additional labor. Sometimes it dropped to the ground while she was standing (6 cases), but usually the mother was lying down (20 of 28 births).

As soon as the calf is out, the cow gets up immediately, sometimes only to her knees, turns to the calf, and begins consuming the bluish membranes of the amnion. Then she licks the calf, especially the face, neck, shoulder, and rump, and continues doing so intermittently for the first few hours, until the calf is nearly dry. Other cows are also stimulated to eat the membranes and may occasionally join in this operation. Once the calf is up, the mother may lick or bite at its umbilical cord, but usually 7–12 cm (2–5 in.) of the cord are left, which shrivels and blackens within a day or so and falls off within one to two weeks. The calf's age can be judged with fair accuracy from the condition of the umbilical cord.

The time from parturition to the first struggles of the calf is quite variable. The calf may "wake up" as soon as its head emerges, shaking its head and moving its ears, or, once free, it may lie motionless on the ground for several minutes. The time required to gain its feet is similarly variable but depends to some extent on the mother: if she is nervous and begins to move away before her calf stands, she urges it to follow by nosing it and calling. The call is a rather faint, abrupt "huh" repeated once or twice a second when urgent, and the calf responds with the same sound in a higher key. When I was close enough to hear the exchange, the calf began bleating almost as soon as it first raised its head.

Most calves begin struggling to rise within three minutes and the majority are able to stay on their feet within another three minutes (see fig. 11.2g). Its first few attempts end in falling, due to a combination of imperfect coordination and weakness—of the front legs in particular. In a sample of 36 timed intervals, 19 calves were afoot within 5 min., 11 within 6–9 min., and 6 took 10–12 min. The average was about 7 min. Once up, the calf is able to gallop beside its mother; in fact, galloping is easier than walking until it gains its balance. For the first hour or so it is constantly knocking against its mother, stumbling over her if she lies down, and breaking into a reeling run like a drunk about to fall headlong—although it seldom falls after the first few minutes.

Unless its mother moves away for some reason, the calf begins seeking the udder as soon as it is afoot. More often than not, it starts at the front and works back, nosing its way up the legs, then along the belly until finally brought up against the back legs. The mother may or may not direct it by nudging or licking its rump but usually gives little assistance. Indeed, some mothers, especially two-year-olds calving for the first time, acted evasively, circling away and even hooking at the calf gently, before finally permitting suckling.

It is unusual for the calf to suckle for more than a few seconds after first finding the teat, and the whole seeking process is repeated numerous times during the first hours before the calf learns to contact the udder by raising its muzzle. Thirteen calves took an average of 9 minutes to get down to proper suckling, which could be distinguished by rapid head-bobbing (but very little of the tail-wagging seen in the young of other ruminants). Within the first day the calf learns to bring down milk by jabbing the udder with its nose, later with enough force to unbalance its mother.

For the first day at least, the mother permits the calf to suckle at will. Thereafter, in 20 cases I timed, the duration was limited to an average of 75 sec. (max. 140 sec.), after which the cow walked away. The calf was evidently never quite satisfied, but after one or two attempts to resume suckling while she was in motion, it would give up for an hour or more. Limitation of feeding in this way could be adapted to maintaining the calf's following response, as well as stimulating early experimental grazing. The number of times calves nurse per day varies, since suckling is the typical aftermath of any disturbance to a herd and of separation followed by reunion of mother and calf.

FORMATION OF THE MATERNAL-INFANT BOND

Wildebeest calves display a well-developed following response as soon as they become mobile. When its mother moves away, a normally responsive calf stays close beside and often in contact with her. However, the following response is indiscriminate until a calf becomes imprinted on its mother. Meanwhile, newborns are irresistibly drawn to any large, moving object, especially to one passing close by and moving away. So long as the mother is the nearest object, this response is adaptive. But an untended calf will approach a lion, hyena, man, or vehicle nearly or quite as readily as another wildebeest. Separation at this stage is therefore particularly risky.

The whole responsibility for avoiding mix-ups falls on the mother—often a trying task in large nursery herds. Recognition of mothers evidently begins with nursing. Consequently, the sooner a calf finds the udder, the sooner imprinting will occur. Anything that delays nursing, such as a disturbance that causes the mother to lead the calf away, may increase the time a calf follows unselectively. Scent precedes voice recognition, or maybe sucking serves to reinforce both the mother's scent and her voice. For mothers, scent is almost certainly the first—and at first the only—sure means of identification. This is clear when a cow is looking for a lost calf. While moving through the concentration in obvious distress, giving a louder version of the calf-call interspersed with drawn-out, plaintive moos, she stops and sniffs heads, shoulders, and rumps of calves encountered on the way. As soon as she smells her own, the distraught mother calms down and stands still or grazes, while the calf nurses.

As mothers recognize the scent of their own offspring within minutes of birth, perhaps licking the newborn, which is only seen in the first few hours, serves this essential function, as is known in goats and sheep.[4] One of the few cases of mistaken identity I saw involved a lost calf that lay touching the flank of a cow in the final throes of labor. When she turned to examine her newly expelled offspring, there was the other calf right beside it. It must have shared the scent of her own calf and/or amniotic fluids, because she allowed them both to suck—one of the few times I've seen that happen. Nevertheless, it became clear during several hours of observation that she could tell the difference and was primarily oriented toward her own calf. When the orphan went off after other wildebeest, she made no effort to bring it back, but merely tolerated it when it came back on its own.

Intolerance of cows toward newborn calves other than their own is an essential feature of wildebeest reproduction (fig. 11.2m). Cows rebuff mistaken attempts to follow or nurse, pushing the newborn away and hooking at it if it persists. This hostile response corresponds closely to the stage during which calves are unable to distinguish their own mothers, thereby helping to minimize mix-ups. Mothers with their own new calves are the most intolerant, and they display corresponding diligence in keeping next to their own offspring. The instant her newborn moves away, the mother follows, and if the calf has attached itself to another wildebeest, she forces her way between them. She may even threaten one that fails to move out of the way; I have seen the rare fight between cows when one was apparently showing an interest in the mother's calf.

Reported instances of wildebeest twins can be accounted for in this way, also by the more common, purely temporary attachment of lost calves to particular cows in a herd. For lost calves will follow their kind (other kinds too; fig. 11.3) and keep trying to find a mother until they are either hunted down or starve. Quite often, calves that appear lost eventually reunite with their mothers. But for those truly lost chances of adoption are virtually nil.

TIMING OF THE AFTERBIRTH

In many mammals the afterbirth is expelled soon after birth. Not so in the wildebeest. Observations of dozens of newborn calves indicate a delay of at least three hours. (R. Hoare, pers. comm., recorded a delay of 5–6 hours following several births during the February 2012 calving

FIGURE 11.3. Anna approached by orphaned calf.

season.) By then calves are dry and well coordinated and, though lacking endurance, are able to gallop beside their mothers at a good pace. The delay is well adapted to the wildebeest's needs, for scavengers are so numerous on the calving grounds, so quick to spot and claim afterbirths, that the rightful owner seldom gets the opportunity to consume it in peace, a process that may take a cow more than ten minutes. Almost invariably, vultures are the first to arrive, often within the first minute (fig. 11.4); birds constantly soaring over the areas where wildebeest are concentrated, beginning within one or two hours of sunrise, make an efficient spotter network. Others wait on the ground, often grouped at sites where they had previously consumed afterbirths, then flap along at low elevation to sites where other birds are descending. White-backed vultures are by far the most numerous, with Rüppel's griffons next most common. Lappet-faced vultures, the largest species, do not compete.

Although jackals, hyenas, and lions all use descending vultures to locate kills, jackals are the only carnivores that compete with the vultures for afterbirths. Singly or sometimes in pairs, golden and black-back jackals prowl and loaf on the calving grounds. When vultures drop from the sky, jackals converge on the spot at a dead run and dash in among the milling birds, causing them to scatter. But the vultures usually succeed in bolting most of the placenta before any jackals arrive.

FIGURE 11.4. Vultures descending on an afterbirth.

If the afterbirth was dropped soon after calving, the risks of predation would be greatly increased, as hyenas undoubtedly would also race
to the spot. The descent of vultures at an afterbirth disturbs any nearby
wildebeest. As birds plummet from the sky, the wind rushing through
their spread primaries makes a noise like tearing silk. Having landed,
they converge on the gelatinous placenta with flailing wings, hissing,
and harsh cries. Mothers of new calves move away first and farthest.
But during and after giving birth, mothers attack any avian scavengers
or small carnivores that come near.

CALF DEVELOPMENT IN THE FIRST WEEK

Wildebeest calves are more precocious than any other known ruminant.
The adaptation to a follower-calf strategy included a speed-up in the
development of locomotion. Gnus have almost adult muscle/bone length
ratios at birth, whereas newborn hider calves have only two-thirds the
adult muscle/bone ratio.[5]
    Most of the innate coordinated movements characteristic of wilde-
beest appear within the first week. Calves are already capable of shaking
themselves all over when they first get up. After a day they have the coordination necessary for the delicate operation of scratching the chin or
behind the ear with the back feet. They lie down by first dropping to the
knees; the linked act of rubbing the forehead on the ground follows somewhat later, as does pawing. Calves cavort and prance in play the first day
and quickly acquire the associated bucking, head-shaking, stotting, and

high-stepping alarm trot. They rub heads and drop to their knees to butt in play-fighting, which is also the position normally assumed while nursing after the first week or two. Mounting other calves and attempted mounting of adults appears within the first few days. Before beginning to eat grass, they rest on their sides or curled up. Experimental grazing begins the first week, as the calf moves along head to head with its mother, as though imitating her choice. By the end of the week, they can be seen lying and ruminating like adults. Eating dung, seen repeatedly, may be necessary to acquire the appropriate rumen bacteria.

SOCIALIZATION

Calves display an early and progressive preference for each other's company corresponding to slackening dependence on their mothers. Once a calf is thoroughly familiar with its own parent it begins to spend increasing time with other young, forming a crèche. The association is closest while resting, when numbers of calves sprawl all together. During early morning and late afternoon activity peaks, calves often join in racing around their elders in the antic manner described. Yearlings, subadults, and even mature cows are sometimes inspired to join in.

Association of calves is the prelude to herd orientation. Thus one can see crèches of month-old young that refuse to obey the summons of mothers standing at a distance and calling them, even though they answer the call. Lying fast asleep while their mothers run about frantically calling them is another indication both of growing independence from the mother and of a sense of security from being present in a herd. This herd orientation gives the mother greater freedom, too. Mother-calf distance while grazing is a measure: when a new calf is sleeping, the mother grazes in a tight circle around it, seldom moving more than a dozen meters away. With older young, the mother tends to graze more in a straight line to greater distances, and may leave the herd for short periods.

Occasionally mothers leave sleeping calves for hours and travel considerable distances. This practice came to light while I was following Ngorongoro cows returning from the lake after drinking that were acting as though they had lost calves: running and bawling, sometimes joining forces and inspecting every herd with calves. After going some 3 km in this manner, several searching mothers found their calves—just where they had left them! Evidently they had not memorized their offsprings' locations. Until then, I had assumed that a lone wildebeest calf was a doomed orphan.

MORTALITY IN THE FIRST MONTH

Abortion and Death in Labor

Spontaneous abortion due to poor condition may account for considerable mortality among calves conceived during periods of drought or disease but is apparently unusual otherwise. I saw only two cases prior to the 1964 calving season. Deaths during labor are equally infrequent, and here again, I witnessed only two cases out of some four thousand females that calved. They died similar deaths after hours of hard labor that was excruciating to watch as it reduced them to final exhaustion.

Separations of Mothers and Calves

Despite the great vigilance of new mothers, separation undoubtedly accounts for considerable calf mortality. This is understandably more pronounced in the Serengeti population, where dozens of abandoned calves may be left behind a migrating army, but a number of lost calves were also to be seen on the Ngorongoro calving grounds, as extras in a nursery herd, or sometimes lying by themselves at a distance, too weak or forlorn to keep looking for their mothers.

Territorial bulls must be listed here as a common source of separations and hence possible agents of calf mortality. Although there are no mating opportunities in the calving season, territorial activity may actually increase. Progesterones in the amniotic fluids are sexually stimulating, a fact well known to livestock owners. Attempts to mount cows that have just calved, or other wildebeest, are not uncommon. Conversely, some bulls display intolerance of aggregations, making periodic sweeps of their territories that clear an area of all wildebeest, including cows and calves. Totally heedless of the consequences, bulls do not exempt cows and calves, even those on a calving ground. It is as if they were saying, in effect, "There are too many animals on my property, eating all my grass, so get out, the lot of you!" Likewise, a resident bull intent on driving a yearling out of a nursery herd scatters everyone in his path. The effect on small herds running the territorial gauntlet to join night aggregations is even more disruptive, but the chances of reunion are better for small groups.

Yet territorial bulls also perform a valuable service by segregating bachelor males, thereby reserving the best pastures for females and young. Cutting out yearlings is part of the process; bulls are most likely to single out ones that interfere with their mother's efforts to stay close

FIGURE 11.5. Legion of the lost. Orphaned calves that stay where they got separated, typically at a waterhole or river crossing, come to greet each new wave of migrants. Meanwhile, their mothers have gone on rarely if ever to return.

to her newborn. While defense of calves is not part of the system, territorial bulls will occasionally chase foraging hyenas but rarely with any follow-up. Another, unintended consequence of the territorial network that can be considered beneficial to females and young is alarm snorts given by outlying bulls that spot prowling predators.

The far greater density of the Serengeti population leads to many more mother-calf separations and are an important cause of calf mortality. Most occur during stampedes of dense aggregations, as when wildebeest panic while crossing rivers, passing through dense vegetation, or going to water. Wildebeest cope with this danger by pursuing what I call a reunion strategy: mother and calf immediately drop out of the throng and run back and forth calling to one another. In this way 99 percent reconnect. Success is indicated by the mother suckling a calf that dives for the udder the moment they reconnect. Still, calves that fail to reunite accumulate at waterholes and crossing points. The sight of dozens of lost ones coming to meet each new wave, hours or days after their mothers have gone on their way, is pitiable (fig. 11.5).

Occasional accidents can result in much greater mortality than usual events like those just described. One extraordinary incident occurred in February 1973, during the calving season while the migration was on the short-grass plains. A large aggregation at Ndutu chose to swim across flooded Lake Lagarja, which is usually shallow enough to wade if not completely dry. Many calves became separated from their mothers during the crossing. Having arrived on the far shore, separated cows and calves turned around and went back. This resulted in some three thousand calves, which are weak swimmers, becoming exhausted and drowning.[6]

What a spectacular miscarriage of the reunion strategy! In the same manner, dozens of calves drown during mass crossings of the Mara River, especially in high water (fig. 11.6). But those losses hardly compare to the death rate when wildebeest choose to cross at locations other than their normal, traditional fords, for example, going into the water down a navigable slope only to be confronted by a steep bank on the other side, as I saw, photographed, and filmed during dry-season months spent in the Masai Mara from 2004 to 2007. Figure 6.6 shows wildebeest tumbling down a 10 m embankment and swimming the Mara in 2005. A large aggregation had passed through riverine woodland only to be confronted with a cliff as they reached the riverbank. There was no turning back, as the hundreds coming from behind pushed those in front over the edge. Miraculously, the video footage of the event recorded no deaths (except for a zebra dragged under by a crocodile) or serious injuries as the opposite bank was low. But all of these accidents pale in comparison to the deaths of some ten thousand wildebeest in the Masai Mara in August 2007, recounted in chapter 6.

Calf Predation

Survival of newborn calves while unable to keep up with the herd or outrun predators depends on avoiding detection and on maternal defense. As already noted, calves are most vulnerable the first night. Within a few days they are too fast and strong, and too closely attached to their mothers, for them to be easily run down by hyenas, the main calf predator by dint of numbers and hunting capability. As a rule, hyenas do not undertake to run down calves more than a few days old, unless they are unattached or there is something wrong with them. For the hyena, perhaps it's just conservation of energy: why go to the trouble of chasing a fit calf when a bigger gnu can be pulled down with only a little more effort?

The first line of defense of newborn calves is for the mother to stay out of sight. Herds of females and calves take flight at greater distances than herds with few or no calves, and mothers with new calves have the greatest flight distance. In a large nursery herd, other wildebeest

FIGURE 11.6. A stranding of wildebeest that died crossing the Mara River.

serve to screen the departing mother and infant from view. Mothers also use terrain to their advantage, by moving out of sight around a hillside, for instance. Calves that stay close beside their mothers are also surprisingly inconspicuous. The light-colored beard (not really white) helps to screen the calf from view, especially if it is on the off-side from a predator's perspective. In fact, mothers moving past a predator often deploy this tactic. Calves stick closest when very young, while running in a group, and when alarmed. Disturbed mothers and young keep so close together that it is hard to spot calves even from the air.[7]

New calves that are targeted by foraging hyenas are almost certain to be caught, unless their own mothers succeed in defending them. With their keen eyesight, hyenas have a remarkable ability to single out calves that are even slightly unsteady in their movements. They are particularly attentive when wildebeest are on the move, and a favorite tactic is to lope into a resting or grazing herd, then wait to see if a new or lost calf or crippled adult fails to keep up with the rest. Once a hyena has started to run down a vulnerable calf, it will overtake it within half a kilometer, unless it loses sight of it. At the moment her calf is overtaken, the mother turns and attacks the hyena—or any other predator less formidable than a lion (fig. 11.7). A pursuing hyena is often bowled over and horned, repeatedly if it persists in attempts to grab the calf. This is usually enough to discourage further pursuit, although I never saw any evidence of injury.

FIGURE 11.7. a) A cow pursued by two hyenas. b) She confronts them while her calf keeps running.

FIGURE 11.7. *Continued* c) One on one a mother can protect her calf but rarely when there are two or more.

Strangely enough, hyenas usually put up no resistance when attacked themselves.[8,9] They generally behave submissively when threatened by wildebeest or other large ungulates, often lying still until the aggressors lose interest. It usually doesn't take long. In his study of Serengeti and Ngorongoro hyenas, Kruuk[10] found that maternal defense saved 21 of 73 calves. Another 36 mothers and calves outran their pursuers—these, doubtless older calves—whereas only 6 that hyenas pursued escaped by disappearing into a herd, probably because newborns were soon left behind by a fast-running herd.

Maternal defense is generally ineffective against more than one pursuer. In 39 attempts to defend a calf, the mother succeeded every time against a single hyena (20 cases) and failed every time against more than one (19 observations).[10] In figure 11.7 a cow has stopped two hyenas that overtook her calf. But meanwhile the calf ran on, and a third hyena that had joined the chase continued pursuit and caught it. Usually maternal defense ceases with the calf's death. When it stops struggling and calling, the mother's flight and herding instincts retake control.

FIGURE 11.8 Wild dogs easily run down new calves.

Defense is still more futile against a pack of wild dogs, which capture calves with ease (fig. 11.8).[9,11] But for every rule, it seems, there are exceptions. The *Wild on Tape* program of September 23, 2008, aired on the National Geographic Channel, includes film footage by Hugo van Lawick showing Serengeti wild dogs failing to overcome a mother's defense of her calf. There was something miraculous in its survival, as different members of the pack of four dogs repeatedly caught and once even dragged it along by the throat. Yet each time, the mother intervened before it was too late. It also helped that the calf stayed as close to her as possible. Finally, inexplicably, the dogs just gave up and stood around the cow and calf, panting but not exhausted. The two gnus then went on their way, the mother in the high-stepping alarm trot, the calf looking none the worse for wear. With hundreds of other calves on the plain, surely the dogs could afford to let one get away.

During the follow-up study of the calving season in Ngorongoro in 1973, birth and survival of young in small herds was a particular focus. We (Runi and I, together with Kathryn Fuller the first three weeks) made daily classifications of females in resident herds between January 7 and February 27. A sample of known individuals and herds was kept under observation through resightings of animals with distinctive natural or artificial marks. (Artificial marks consisted of red paint in darts with the needles removed shot from a CO2-powered Cap-Shur gun. It was a failed effort, as the blotches on wildebeest hides wore off within a week or so.) Even so, we were able to monitor 25 known individuals in nine different herds containing an average of 71 cows for periods ranging from eleven to forty-four days (average of 13 resightings). The status of females in aggregations was assessed periodically for purposes of comparison.[1]

There were very few calves to be seen during the first week. And to our surprise, the hyenas seemed oblivious of them. Time and again we saw foraging hyenas ignore clearly visible new calves. Within several days we saw a dozen or more cows with early calves scattered over the Crater floor. Then one morning we saw none. Evidently wildebeest calves had been off their menu for so many months that hyenas had put them out of mind. Now the Gestalt of a gnu calf had been reprogrammed. Consequently, very few calves born before the main peak began around February 12 survived. Meanwhile, hundreds of calves had been born in the resident small herds.

Predators caught most calves born before the main birth peak. By then, hardly 3 percent of all the pregnant cows in Ngorongoro had calved, and it was unusual to see any calves in an aggregation. In effect, calves in the small herds were providing a steady protein supplement for Ngorongoro's 470 hyenas. The absence of large numbers of calves to glut the market and provide cover for newborns meant that even calves well past the feeble stage could be kept in sight and run down. The tendency of mothers with newborns to be the first to run from predators compounded the problem, as they tended to flee from one herd and run in plain view to another, positively inviting pursuit.

This failure of a strategy that is adaptive in calving ground aggregations is comparable to the failure of the reunion strategy under unusual conditions. The increased risks of calving in small herds is a strong argument that the follower-calf strategy evolved in migratory populations.

To our surprise, calving among the resident herds peaked at different times in different parts of the Crater. Their schedules seemed to reflect the rainfall and vegetation gradient inside Ngorongoro: well watered on the east side in the path of the northeast trade wind, becoming increasingly arid toward the center and west sides. Herds on the east side calved first, beginning the second week in January. Calving began on the central plain about the same time but only peaked the third week; it was over by the first week in February, just as calving in small herds rose to a peak in the hilly northwest area.

Birth peaks also varied between calving grounds. Mobile aggregations that remain on opposite sides of Ngorongoro for months at a time—as often happens—would also be likely to reproduce on a somewhat different schedule, reflecting differences in rainfall and pasture quality prior to the rut. But lack of synchrony between aggregations containing thousands of animals would be apt to have less serious consequences, in terms of calf survival, than lack of synchrony between resident subpopulations.

Despite unsynchronized birth peaks in Ngorongoro subpopulations, once the peak started within a subpopulation it continued for the usual three weeks, with the majority of calves born during the middle week. However, calving prior to the peak acted to prolong reproduction in the small herds and increase mortality.

Survival rates in the small herds reached a maximum when births on the calving grounds peaked, flooding the Crater with some 3,700 offspring, in a population estimated at $12,000^2$. The overall average in a final sample of small herds taken on February 22, after 97 percent of pregnant females had calved, was slightly over 50 percent (sample of 226 postpartum cows). Survival of calves born in aggregations, starting at only 20 percent the first week, quickly rose as their numbers multiplied and reached 85 percent before the peak ended. The resident small herds benefited by calving at the same time: 70 percent survived, compared to only 28 percent on the central plain where calving finished before it started on the calving grounds.

But the 1973 survival rate of nearly 85 percent in Ngorongoro aggregations was exceptional. The mean survival rate after five Ngorongoro calving seasons (1965, 1969, 1973, 1975, 1977) was 74.9 percent in aggregations and 41.1 percent in resident small herds[2]. A calf:female sample I took in February–March 1997 was about the same, but one I took in March 1999 indicated a considerably lower survival rate (62 percent). Although these samples included both aggregations and small herds, the calf:cow sample I took of Serengeti aggregations in March 1999 was barely over 50 percent.

Serengeti postcalving samples Mduma[12] took in March 1993 and 1994 yielded calf:female ratios of 78.3 percent and 28.3 percent, respectively, illustrating great variability of survival rates from one year to the next. At the end of the year the ratios converged at about 20 percent, indicating that 80 percent of the year's calves died before they turned one year old. No doubt most of this mortality reflects nutritional stress in the dry season; thus the low calf:cow ratio of the 1994 calving season shows the impact of the unusually severe 1993 dry season on the 1994 calf crop. But some of the generally higher mortality of Serengeti calves could result from deaths during or following stampedes and accidents. Yet yearling:cow ratios in Ngorongoro and Serengeti were close at 38.5 percent and 37.8 percent. Mduma calculated that yearling recruitment in 1993 and 1994 amounted to 6.2 percent and 7.3 percent of the Serengeti population of 1.2 million. The effects of the 1993 drought on the 1994 Serengeti calf survival rate was not felt in Ngorongoro. But a disastrous Ngorongoro drought in 2000, which did not affect the Serengeti, led to the near-total mortality of the 2001 calf crop, as virtually no yearlings were to be seen in March 2002.[3]

HOW PREDATION SHAPES THE CALVING SEASON

By eliminating nearly all calves born ahead of the birth peak, predators effectively synchronize the onset of the season, mainly through culling by the spotted hyena. Consequently, the birth peak begins abruptly, reaches a maximum around the middle of the three-week peak, then tails off, with up to 20 percent of the year's calf crop produced over the next five months. Calves born after the peak can benefit from the two-month grace period before the tan color of the newborn changes to the dark color of adults. Calves born later are again conspicuous. Data on their survival rate are lacking, but all those I have seen were single and stood out among the thousands in an aggregation of gnus dressed in the species' distinctive uniform.

Considering all the different factors that impinge on wildebeest reproductive success, variations in calf survival are only to be expected. Practically everything that affects reproduction depends on rainfall amount and distribution, beginning with range conditions and including wildebeest distribution and grouping patterns during the calving season, fecundity, birth synchrony, and condition of the mother and offspring at birth. All of these factors influence predation and calf survival.

REFERENCES

1. Estes 1976.
2. Estes and Estes 1979.
3. Estes, Atwood, and Estes 2006.
4. Hersher, Richmond, and Moore 1963.
5. Grand 1991.

6. Sinclair 1979a.
7. Talbot and Talbot 1963b.
8. Estes 1967b.
9. Estes and Goddard 1967.
10. Kruuk 1972.
11. Schaller 1972.
12. Mduma 1996.

CHAPTER 12

Serengeti Shall Not Die?

*Africa's Most Iconic World Heritage Site
under Siege*

That the Serengeti migratory ecosystem remains largely intact would seem to defy the odds, for all other African migratory ecosystems (apart from the floodplains of South Sudan, where the kob and tiang migrations miraculously survived the civil war) have been severely disrupted. Even the Serengeti ecosystem lost an estimated 40 percent of its original area (approx. 30,143 sq. km in 1910) by the 1990s.[1]

I have dwelled on the characteristics that set the Serengeti apart from all other savanna ecosystems in the preceding chapters. In this final chapter I want to enumerate the challenges to survival of this most iconic World Heritage Site (WHS) and International Biosphere Reserve (IBR) in the twenty-first century.

Only Serengeti National Park and the Ngorongoro Conservation Area carry the WHS and IBR designations. The other parts of the ecosystem, described below, are less well protected.

In Tanzania:

Game Reserves—Maswa (2,200 sq. km), Ikorongo (3,000 sq. km, 60,274 ha), Grumeti (2,000 sq. km) and Kijereshi (65.7 sq. km)

Game Controlled Areas—Loliondo (4,000 sq. km), Speke Gulf (300 sq. km)

Open Areas—Makao (1,330 sq. km), Ikoma (600 sq. km)

Community-based conservation schemes, including Ikona Wildlife Management Area (WMA)

In Kenya:

Masai Mara National Reserve (1,500 sq. km)

Mara group ranches—Masai lands abutting the Masai Mara Reserve. The four main ranches are Koyiaki, Lemek/Ol Chorro, Olkinyai, and Siana.

The Isiria plateau (drought refuge, unprotected)

Overall, 20 percent of the ecosystem is unprotected.

DIFFERENCES IN THE LAND TENURE SYSTEMS OF TANZANIA, KENYA, AND UGANDA

To appreciate the challenges to the Serengeti ecosystem, you need to understand the different approaches to land stewardship in each country, which reflect dissimilarities in political systems and governance. All three are republics, but two have socialist backgrounds, and one, Kenya, is arguably the most capitalistic African nation.

In precolonial Africa, the underlying principles of land relations were rooted in social and political customs, which have underpinned and still influence ongoing changes in the tenure systems. Traditional systems of local land tenure clearly defined individual or family rights to some types of land use, for example, agricultural and pastoral lands, as well as common property resources. Rights were conferred on the basis of accepted group membership while maintaining a degree of group control or supervision over land affairs.

Tanzania

Just prior to independence in 1961, the British colonial government attempted to introduce the concept of freehold landownership, but the proposal was rejected by TANU, the political party of President Julius Nyerere, who was intent on making Tanganyika a model of African socialism. The Arusha Declaration of 1967 made Ujamaa, or familial unity, socialist development the law of the land. The entire body of land in Tanzania was declared "public lands." Customary land rights and chieftainships were abolished and district and village governance systems established. Rights over the land were placed under the control and direction of the president of the United Republic, and those rights could not be disposed of without the president's consent.

The next step was the Villagization Program of 1974, which entailed the relocation of an estimated 75 percent of the population from scattered homesteads and smallholdings to communal *(ujamaa)* villages of 2,000–4,000 residents. The benign goal was to improve social welfare, but when necessary, relocations were enforced. Village councils were given responsibility for land allocation and management. By implementing this large-scale resettlement program, the central government essentially collectivized all forms of local productive capacity. In addition, the government nationalized and owned all of the major enterprises of national importance. To the extent that Ujamaa removed hunting pressure on wildlife over a wide area, the ecologist Alan Rodgers considered that large mammal populations probably benefited (N. Stronach, pers. comm.).

Nyerere's socialist experiment resulted in severe economic decline. Collectivization, intended to increase productivity, didn't work: independent farms produced twice as much as the collectives. By the time Nyerere stepped down in 1985, Tanzania had become one of Africa's poorest countries, dependent on international aid.

Tanzania had reached rock bottom by the time I returned in 1978 to spend two and a half years at the Serengeti Research Institute, accompanied by my wife and two children aged five and seven. What an amazing contrast to the 1960s, when I lived in relatively high style in Ngorongoro Crater. Now there were virtually no imports, and shortages existed for even such basic things as butter, bread, fuel, and toilet paper! When shopping, the reply to practically every item on the list was *hamna*, KiSwahili for "there is not." I still keep a worn-out T-shirt bearing the motto *Hamna Tanzania*. A major reason for the ultimate Tanzanian economic debacle was the 1979 invasion of Uganda to get rid of the abominable Idi Amin and restore Nyerere's crony, Comrade Milton Apollo Obote, to power.

To live anything approaching a normal life in the Serengeti, we made supply runs to Nairobi every few months. The Tanzania Immigration officials at the Bologonja border post let us through, providing us with shopping lists of what to bring back for them. As the international border was closed at the time, there was no checkpoint on the Kenya side (we registered with Immigration upon arrival in Nairobi). Our first stop, eagerly anticipated by all, was Keekorok Lodge, where we feasted on Cadbury chocolate bars.

Neil Stronach, son of Brian Stronach, the well-known Tanzania game warden, who was doing doctoral fire research and managing antipoaching in Serengeti National Park from 1985 to 1987, contributed this vignette on the recruitment of soldiers returning from the Uganda incursion.

> TANAPA inherited as rangers many demobilized soldiers from the army. Because of their reputation as ferocious fighters, the Wakuria were preferred and willing recruits to the army. Being local, most of these Serengeti demobs were Wakuria. They brought a level of gratuitous violence to antipoaching that was effective in the short term but probably detrimental to conservation efforts in the longer term. Interestingly, as related to me by several of these rangers, when Tanzanian soldiers fell in the Uganda war, their bodies were repatriated by plane with full military honors. Some of those coffins, at least, did not contain bodies but rather automatic and semi-automatic weapons and ammunition which the Wakuria used to good effect to rustle cattle from the unfortunate Masai and to hunt elephants and rhinos in Serengeti. It was a source of considerable amusement among those rangers to think of the coffins loaded with arms being saluted and paraded at the airport.

The thousands of weapons brought home by the troops soon flooded Tanzania. Gangs of bandits armed with AK-47s began hijacking buses and shooting up shops, offices, and police stations. The Wakuria started to shoot up both the Serengeti wildlife and the rangers, and to rob tourists. Serengeti NP could do little to stop them. By 1986, as the result of drastically reduced tourism and operating budgets, only a single vehicle was available for antipoaching patrols.[2]

"So began a period of murder and mayhem that was to last seventeen years—and devastate the Serengeti ecology in the process," writes Tony Sinclair.[3] Sinclair together with other dedicated researchers and their assistants continued their ecological research during most of this time, despite many close shaves.

After President Nyerere left power in 1985, succeeding governments began to chart new directions for Tanzania's economy and society. A new approach to property rights and resource governance was obviously needed, and in the early 1990s the Presidential Commission of Inquiry regarding Land Matters considered sweeping reforms: a gradual transition to a legal framework that supported private property rights, permitted individualized (rather than collective) control of resources in farming areas, and promoted private investments that utilized Tanzania's natural resources for economic gain. Land legislation enacted since the mid-1990s recognizes long-term occupancy rights to land and allows for land inheritance and transfer.

The Village Land Act (1999a) recognizes the rights of villages to land held collectively under customary law. Holders of customary rights of occupancy may lease and rent their land, subject to any restrictions imposed by the village council (GOT Land Act 1999b). Village land allocations can include rights to grazing land, which are generally shared. The village council may charge annual rent for village land (GOT Village Land Act 1999b).

Nevertheless, the land policy bows to Tanzania's socialist legacy by reaffirming that the president controls all land as trustee on behalf of all citizens, and any property rights granted are only land use rights. Thus rights gained under prior regimes or under customary systems are in reality insecure. All land, including land held under occupancy rights, is vulnerable to expropriation by the government for uses deemed to be in the public interest. According to the USAID Country Profile:[4]

> On a practical level, the authority that remains with village and regional authorities over the country's so-called village lands provides limited opportunities to buy, sell, consolidate or improve those smaller plots of land that are most accessible to smallholder farmers. The process of transferring customary land rights is cumbersome, confusing, and lacking in transparency. Thus, opportunities for efficiency improvements and entrepreneurship among less empowered constituencies have not improved with changes in the law. Although private ownership of land is legally possible, and very important in urban areas, most of Tanzania's farming and herding land remains, as a practical matter, communally owned and controlled.

Land and Wildlife Conservation

The built-in conflict between traditional local land tenure and government ownership of all land is reflected in the measures taken to protect Tanzania's wildlife resources.

Natural resources are broadly governed by more than eight departments, including Wildlife, Forestry, Water (under formulation), Beekeeping, Fisheries, Land, and Environment. The Tanzania Ministry of Natural Resources and Tourism (MNRT) is responsible for conservation of natural and cultural resources as well as development of tourism. Institutions under the ministry's jurisdiction include the following:

National Parks (TANAPA), a parastatal with its own board of trustees, whose core mission is conservation of ecosystems in all areas designated as national parks

Ngorongoro Conservation Area Authority (NCAA)

Forest Service, responsible for protection and management of forest reserves

Wildlife Division, responsible for protection and management of all other protected areas; and oversees hunting

Tanzania Wildlife Research Institute (TAWIRI), established in 1980 with the mandate to carry out and coordinate wildlife research

Other departments involved in wildlife and natural resources are:

Mweka College of African Wildlife Management (CAWM; a.k.a. Mweka), another parastatal organization established in 1963, where many TANAPA rangers and wardens are trained

Pasiansi Wildlife Training Institute, established in 1966 to train Wildlife Division game scouts

Forestry Training Institute—Olmotonyi (FTI), Arusha, established in 1937 to provide two-year training for forest rangers at certificate and diploma level

Bee-keeping Training Institute, offers certificate and diploma courses in bee-keeping

Community Based Conservation Training Centre (CBCTC), established in 1995 to implement the National Wildlife Policy, which emphasizes community participation in the conservation and utilization of wildlife for sustainable development

When a ban on tourist hunting, imposed in 1972, was lifted in 1978, the government gave itself a monopoly on managing hunting blocks by forming the parastatal Tanzania Wildlife Corporation (TAWICO). In practice, though, TAWICO sublet many blocks to private, and mainly European, outfitters. Ongoing corruption brought TAWICO's monopoly to an end in 1984, when nine private outfitters were allocated hunting blocks for periods of up to four years.[5,6] But this parastatal continued to engage in commercial wildlife utilization, including game cropping and the capture of wildlife for zoos, circuses, and the pet trade. A private hunting company assumed management of TAWICO's tourist-hunting operations in 1998. Presently known as Tanzania Adventures Inc., owned by Jack Brittingham, it is one of the leading and most expensive hunting companies, with leases on choice hunting blocks inherited from TAWICO in, for example, Selous and Natron.

It is worth mentioning here that of the three countries whose wildlife policies are under review, Tanzania is the only one that has separate game departments and national parks. Kenya and Uganda merged theirs in the 1970s. A commission that studied Tanzania's wildlife organizations in the late 1990s also proposed consolidation, but the organizations resisted. Reluctance of the Wildlife Division to share income from tourist hunting could have been a consideration. (See why below.)

Oversight of Protected and Unprotected Areas by the Wildlife Division (a.k.a. Game Department)

With responsibility for managing and protecting all lands outside of national parks, the NCA, and forest reserves, the Wildlife Division (WD) is not only responsible for all other protected areas but also for wildlife on Open and unclassified land.

Twenty-five percent of Tanzania's land area is reserved for wildlife protection, the highest percentage of any African nation. But how effectively are habitats outside the core areas of national parks and game reserves protected from human exploitation?

The level of protection afforded these different kinds of reserved areas slides downward from game reserves to open areas.

- *Game Reserves*—Tourist hunting but no settlement permitted.
- *Game Controlled Areas* (GCA)—Settlement, grazing, cultivation, and tourist hunting permitted.
- *Open Areas*—Limited cattle grazing, firewood collection, bee-keeping, tourist hunting and resident hunting, game-cropping and capture (for export to zoos, exotic pet trade) all permitted. Resident hunting is prohibited in all other protected areas.
- *Wildlife Management Areas* (WMAs)—The latest (1998) attempt to conserve wildlife and enable communities to profit from setting aside part of their land for conservation. A WMA is an area agreed to by the minister of wildlife and local village government that is dedicated to biological natural resource conservation.[7] WMAs can be set up within game-controlled areas or open areas but not in national parks, conservation areas, or game reserves. Tourist hunting is their best and in many cases only prospect for being profitable, unless they happen to border a popular protected area (e.g., Masai Mara and Maswa Game Reserves) and can lease land to tour companies.

Open areas and some GCAs may actually exist only on maps. Furthermore, the Wildlife Division has consistently balked at granting WMAs local control, while the hunting industry as represented by TAHOA (Tanzanian Hunting Operators Association), has been vehemently opposed to community-based conservation. Their combined influence has played a major role in delaying the process. Hunting outfitters, accustomed to renting government hunting concessions for minimal fees, have been unwilling to openly compete for community-controlled hunting concessions. Consequently the devolution process has dragged out for nearly two decades, with only sixteen or so WMAs gazetted.[8] Interestingly, virtually all these WMAs depend on the support of nongovernmental organizations (NGOs).

Sport Hunting. One of the Wildlife Division's primary functions is regulating tourist hunting, which is a permitted activity except in parks, forest reserves, and the NCA. Tourist hunting of African big game brings in as much money as photographic tourism, which amounts to some 16 percent of Tanzania's foreign exchange. The ratio of tourists who come to see the wildlife and hunters who come to shoot it is many hundreds to one—arguably mirroring the difference between the U.S. middle class and the millionaire top one percent! The Wildlife Division earns most of its income (60 percent) from license fees levied for each species available in hunting blocks. These are leased for five years at a time to outfitters that employ licensed professional hunters (PHs) to guide the tourist sportsmen. Each year the WD sets the quotas in the hunting blocks, and hunting companies have to generate revenue of at least 40 percent of the value of the quota for each species. Failing to do so, the outfitter is required to make a top-up payment to the WD to meet the 40 percent minimum. The outfitter is further required to contribute to antipoaching, road construction, and community development. Outfitters also help to build schools, drill water wells, bring in doctor(s)/nurses, and provide protein to locals by donating hunted meat, along with employing the local people to act as company trackers, skinners, cooks, and cleaning ladies. (How thoroughly these requirements are fulfilled may vary.)

Clients have the choice of signing up for safari packages lasting 10 or 21 days and selecting the species they want to shoot ahead of time, so each package has a specified number of animal species to be hunted during the July–December hunting season. A safari can last a shorter or longer time, but clients must pay either a 10-day or a 21-day license and the fees associated with each. The most sought-after big game, headed by a male lion,

FIGURE 12.1 (TABLE)   ESTIMATED COST OF 21-DAY HUNTING SAFARI TO
NATRON HUNTING BLOCK IN 2012 WITH TANZANIA ADVENTURES INC. IN ITS
LAKE NATRON GAME CONTROLLED AREA HUNTING BLOCK IN 2012

Hunting Client @ $3,060/day	Subtotal	$64,260
Government Fees (estimates only & subject to change)		
Hunting license @ $1250		1,250
Wildlife Conservation Fee @ $150		3,150
Block fee - hunting client @ $500/day		10,500
1 21-day license @500		500
Trophy Handling Fee license @ $2,000		2,000
Trophy Fees (Estimate)		
Deposit towards estimated trophy fees		20,000
Area Transfers (Estimate) Date Airstrip/Location		
To: 1 road transfer Arusha to Natron @ $350		350
From: 1 road transfer Natron to Arusha @ $350		350
Miscellaneous		
Additional car/driver (bait truck) $150 per day		150
Hotel accommodations in Dar es Salaam (pre/post-safari) to be determined, based on international flight schedule		
Camp satellite phone/email use $1.00/min, $8.00/Mb sent/rec'd as used		
Firearm rental (rifle or shotgun) @$250/week or $30/day per gun		
Premium ammo purchase @$11 per round.		
Subtotal govt fees, est. Trophy fees, charters, & misc.		$38,500.00
Pre-safari total including trophy fee deposit		$102,760.00

can only be hunted on the long license. Here's what you could expect to pay in $US on a 21-day hunt offered by Tanzania Adventures Inc. in its Lake Natron Game Controlled Area hunting block in 2012 (fig. 12.1).

The actual trophy fees paid by the client depend on which and how many of the available species are bagged. Below is a sample of what is on offer in the Natron hunting block, in descending price:

Lion $4,900, leopard $3,500, fringe-eared oryx $2,800, lesser kudu $2,600, gerenuk $2,500, buffalo $2,400, crocodile $1,950, eland $1,700, klipspringer $1,600, zebra $1,350, wildebeest $1,250, ostrich $1,200, Coke's hartebeest $650, bushbuck $600, Grant's gazelle, bushpig, spotted or striped hyena $450, wildcat $400, impala $390 . . . baboon $110.

An antipoaching and conservation fee is levied on each kill, ranging from $3,100 for a lion to $90 for a vervet monkey. So if you bag a lion your total trophy fee will amount to $8,000.

The daily fees are paid to the hunting company; the Wildlife Division collects the trophy and concession fees. Average income to the WD per hunting client is around US$7,000.[8]

When all the charges hunting clients must pay are added up, it is amazing that so many choose Tanzania over South Africa, Zimbabwe, or Namibia, where a three-week hunting safari may cost half as much. (Botswana and Zambia recently closed hunting.) But one big difference: unlike nearly all hunting reserves in southern Africa, Tanzania's hunting blocks are still unfenced and therefore perceived to be wilder.

"The biggest reason people hunt Tanzania," says Olivia Opre (pers. comm.) of World of Hunting Adventure, "is the quality of the big five (mainly lion & buffalo) and the ability to collect/hunt an enormous diversity of game. . . . Tanzania also conjures up thoughts of the 'classic African safaris' that Ruark, Hemingway or Roosevelt went on and depicted in the books that many hunters have read." (Masailand has gerenuk, lesser kudu, Thomson's and Grant's gazelle, available only there and in Ethiopia, where they're classified differently according to Safari Club International.)

The difference between the cost of a hunting safari and a tour through Tanzania's protected parks and reserves is mind-boggling. A 16-day guided tour on the Northern Circuit, starting with Tarangire and including Serengeti NP, costs $7,000 to $10,000, plus travel to and from Tanzania.

As wildlife populations have declined in many areas of Tanzania, it is predictable that animal rights activists would try to lay the blame on sport hunting. Over 75 percent of the protected areas in Tanzania were originally set aside for trophy hunting. Approximately 20,500 hunting days are sold annually to 1,370 clients in 180 hunting blocks, generating a gross income for the industry of over US$27 million from daily rates. But according to Baldus and Cauldwell,[8] "There is no evidence that the regulated tourist hunting industry has contributed to the general decline of wildlife populations, but there is plenty of evidence that the presence of a regulated hunting industry contributes significantly to reducing the illegal activities of poachers and provides an economic incentive to protect vast areas" (26).

Baldus and Cauldwell[8] warned that restrictions on hunting imposed by CITES, the revised U.S. Endangered Species Act, and the negative influences of antihunting lobbies are the forces that are destructive and dangerous to the Tanzanian economy. "Pending reform," they wrote, "the Wildlife Division is extremely vulnerable to these negative forces" (43).

During the time (1987–2002) when the German Technical Cooperation Agency (GTZ) managed the Selous Conservation Programme, Rolf Baldus was able to win support for community-based natural resource management in the 15,000 sq. km of buffer zones around the 50,000 sq. km Selous Game Reserve (SGR) by offering a fair share of the profits derived from tourist hunting. Suddenly wildlife became an economically valuable resource. Tourist hunting provided around 90 percent of all SGR's retained revenue.

But all that changed after the GTZ contract ended. Lax law enforcement accompanied by renewed heavy poaching (currently decimating Selous's elephant population), awarding hunting blocks to crony hunting outfitters in a nontransparent way without competitive bidding, foot-dragging the creation of WMAs for two decades, and the hunting lobby's unwillingness to compete for concessions in rural communities stifled community-based conservation.

In 2004, the Tanzania Development Partners Group, representing all donors of development aid, strongly criticized poor governance in the conduct of tourist hunting by WD and the lack of benefit-sharing with communities. Considering that the Wildlife Division would be destitute without hunting fees, its unwillingness to lose them to local communities is understandable. Merge with TANAPA? Forget about it! Indeed, the recent Wildlife Conservation (Non-Consumptive Wildlife Utilization) Regulations 2007 have concentrated all management powers and revenues centrally, instead of devolving such powers to, and sharing benefits with, local communities. Endless delays have fueled speculation that opportunities for private gain by senior public officials and officers of TAHOA underlie such delays.[5,9] Then, too, there is always the temptation to subdivide hunting blocks and raise the quotas that hunting companies must pay.

In 2009 Baldus published *A Practical Summary of Experiences after Three Decades of Community-based Wildlife Conservation in Africa: "What Are the Lessons Learnt?,"* from which I have taken the following excerpt.

> Community Based Conservation (CBC) . . . is based on the assumption that the interface between rural people and wildlife is dependent upon incentives. CBC aims to achieve results in conservation and rural development by creating economic incentives and suitable proprietorship. The necessary institutional reforms therefore had to be closely connected with the devolution and decentralisation of formerly centralised decision-making. Such approaches link biodiversity conservation with rural socio-economic development in a

way that benefits local communities. The aim was conservation, but at the same time improved rural development and combating poverty. The guiding assumption was that people who benefit directly from the use of natural resources will develop an interest to protect these resources and to limit their use to sustainable levels.

There is a third approach to convince rural people to commit themselves to conservation. It consists of appealing to them to protect wildlife, as it will otherwise go extinct. This has been widely attempted by animal preservationists, yet there is little evidence that it has ever worked in reality. Wildlife education and the reference to the intrinsic values of animals might enlighten the rural dwellers and might encourage them to think about animals more positively and more often. Positive attitudes towards game do, however, not necessarily lead to positive conservation action. They may influence, but will not determine people's concrete dealings with such animals, for example whether or not they will hunt them for food or whether they will tolerate animals on their land. Primarily livelihood aspects and economic considerations influence such decisions. Rural dwellers cannot be convinced that wildlife is nice to look at and even worth preserving, when this same wildlife threatens their livelihood, eats their crops and even kills family and community members. (12)

Rundown of Serengeti Protected Areas

Over 80 percent of the Serengeti ecosystem is legally protected through a network of protected areas (PAs) (see map 5.2), although the management regimes of some are too weak to guarantee their effective protection. The current population in the seven districts to the west of the park is over two million (more people than the total population of Botswana), with an annual growth rate exceeding the national average of 2.9 percent.[10] It is a multiethnic region with over twenty tribes[11] earning their living through crop production, livestock husbandry, charcoal burning, hunting, and mining. The consequent need for more land and food is putting increasing pressures on protected areas.

How Reliable Is Wildlife Protection in Game Reserves?

The Ikorongo, Grumeti, and Kijereshi Game Controlled Areas were upgraded to game reserves in 1994. Although their protected status is second only to national parks, game reserves are generally less strictly protected than parks. Intrusion of livestock, removal of forage and firewood, honey harvesting, fishing, and other activities that surrounding settlements pursued until their land was expropriated are often tolerated. Poaching for meat is a serious problem in game reserves. Game reserves are also less resistant to illegal settlers from neighboring communities who argue their need for more land should be considered more important than preserving it only for wild animals.

Maswa Game Reserve

Maswa Game Reserve is a case in point. Poaching has long been a serious problem. A total of 32,613 wire snares were collected in Maswa Game Reserve between 1989/90 and 1998/99, an average of 3,261 snares a year.[12] However, illegal settlement is arguably more serious, as it has led to boundary realignment three times, causing about 15 percent loss of the original area.[13] Most recently, in January 2011, the district commissioner served notice that settlements inside the reserve would have to be vacated within six months, on the grounds that the five hundred or so pastoralists with more than 213,000 cattle, who had been living there for more than five decades, were degrading the habitat. The pastoralists, although belonging to different tribes (Sukuma, Masai, Barabaig, and Taturu), refused to pull up stakes. They complained that, among other things, the eviction notice was handed down without prior consultation and no alternative area where the displaced people could relocate was offered.

They had reason to be suspicious of government motives because of past evictions of pastoralists from traditional grazing lands. The comments of one of them sums up the prevailing attitude. "As pastoralists, we'll not leave this area. These people [the government] have their own agenda—to grab our land for an unknown reason," he said. "These WMAs are not friendly to the pastoralists. They are pro-investors and I don't know why the government is engineering these kinds of projects that favor aliens on our land?"[14]

Previous disputes involving settlement inside the Maswa Reserve were resolved by simply excising the disputed area. Faced with a difficult political choice, the WD and the district commissioner elected to give in to the settlers. On September 4, 2003, Tanzania's *Guardian* newspaper reported that the government had declared a new boundary for Maswa Game Reserve to end the long-standing conflict between the Wildlife Division and people who had unlawfully settled inside it. The government gave 35 sq. km of the encroached reserve land to the illegal settlers because it "didn't want to disturb the wananchi (people of the land) and it avoided inconveniences of resettlement."

Two of Tanzania's most highly respected outfitters have pioneered community-based conservation in the Maswa area. The Robin Hurt Hunting Company and the Friedkin Conservation Fund have for years backed up the Wildlife Division in controlling poaching, cattle encroachment, fires, and illegal dwellers in their concession areas bordering and inside the reserve.

That may be grounds enough to fuel the resentment of pastoralists and others who are kept out. However, the rangers who patrol these areas are recruited and trained locally. Both companies have established foundations based on the premise that for wildlife to survive in a changing Africa, it must be a competitive form of land use, benefiting human communities.

Dedication to poverty alleviation among village communities and sustainable development within communities is the mission of the Robin Hurt Wildlife Foundation. The Friedkin Conservation Fund (FCF) has recently established the Mwiba Game Reserve on land leased near Makao Village, which is surrounded by four conservation areas. FCF is establishing its first Community Center in Makao, to be built and run using funds and people from both Makao and Mwiba. Both companies also have CBC projects in their other Tanzania concessions: FCF's holdings total more than six million acres that are dedicated to conserving Tanzania's protected areas, including the wildlife, wilderness, and local stakeholders. Its Lake Natron Game Reserve concession is unique in that it is located exclusively on village and tribal land.

What are the chances of a different outcome this time? After repeated boundary changes the squatters know now that they will always come out the winners.

Grumeti and Ikorongo Game Reserves

It was a different story in these two protected areas bordering the western salient of the park (see map 5.2). Formerly, both were classified as "Game Controlled Areas" in which settlement and almost everything else allowed in open areas and on unprotected lands was permitted. In the Grumeti GCA, I recall a tiny village on the all-weather road that crossed the Sabora Plain between Ft. Ikoma and Musoma, as well as scattered small settlements protected by euphorbia hedges on hilltops viewed while driving cross-country from Kirawira back to Seronera.

In 1994 the status of these two GCAs was upgraded to game reserves. That meant the settlements had to be removed and the other uses permitted in GCAs prohibited, except in the 600 ha Ikoma Open Area. The army and police were called in to move and resettle the residents. The Grumeti and Ikorongo Reserves were then leased to VIP Safaris, a Tanzania hunting company that contracted with independent professional hunters to bring their clients on safari hunts, which had to be accompanied by Game Department scouts.

VIP managed to secure an enclave of 5,000 ha on and surrounding Sasakwa Hill, with a hundred-year lease granted by the nearby village of Makandusi. (The research on reproductive physiology of the Serengeti wildebeest described in chapter 10 was carried out between 2001 and 2004 at the foot of Sasakwa Hill, by arrangement with VIP Safaris.) After Paul Tudor Jones, the American billionaire and conservationist, bought out VIP in 2002, the high-end Sasakwa Lodge was built on the hill, creating hundreds of jobs for residents of Makundusi. (See Singita Grumeti Reserves website for more information.)

Before and for some time after VIP took over, poaching was uncontrolled in the two reserves. At night, looking out at the superb view from VIP's hilltop tented camp, you could see headlights crisscrossing the Sabora Plain, as poachers hunted for the pot and, in time, for the bushmeat market. Resident ungulates were in low numbers, and shooting from vehicles had made the animals flee at sight of a car half a kilometer away. Only wildebeest resident inside the park could be approached close enough to dart and radio-collar. Amazingly, a megaherd of some five hundred topi managed to survive the slaughter by keeping a good kilometer away from vehicles.

After Jones bought out VIP in 2002, his rangers took on the poachers by conducting foot patrols day and night as in former times instead of patrolling in vehicles as in Serengeti NP. Resident game populations soon rebounded and gradually ceased to regard autos as predators. Recruiting poachers as game guards, a common practice in African parks and reserves, increased the efficiency of these antipoaching efforts.

The Ikoma Open Area and the Ikona Community Wildlife Management Area

The Serengeti Regional Conservation Strategy (SRCS) undertook to establish a WMA in the Ikoma Open Area between the park and the Grumeti and Ikorongo Reserves. Funded by Norwegian Aid and the Frankfurt Zoological Society (FZS), it was an attempt by the park to gain the support of six communities near Fort Ikoma through sustainable use of the wildlife resources. But the approach was too top-down to work. For instance, the Wildlife Division shot and delivered pickup loads of game to the villages in the naive belief that it would reduce the incentive to poach. Instead, it may have been seen as a case of do as we say (conservation), not as we do (hunt bushmeat). The SRCS closed down in 2000. Poaching in the open area and in the adjacent protected areas continues to be a major problem.

Loliondo Game Controlled Area

Loliondo is in Masailand, which overlaps both sides of the international boundary. It is savanna with a low population density of pastoral Masai who depend on cattle and tolerate wildlife. Consequently wildlife is more abundant than in surrounding mixed-use areas. The British colonial power set aside 4,000 sq. km of the Loliondo region as a game reserve in 1959 (the same year the Ngorongoro Conservation Area was gazetted). Hunting in the reserve was reserved for European royalty. After Tanganyika became independent in 1961, President Nyerere took upon himself the powers to issue hunting permits for Loliondo, but he never granted any.

Or hardly any. According to N. Stronach (pers. comm., 2013), "In 1972 the renowned game warden, Eric Balson, and Myles Turner guided Prince Bernhard on a hunting safari in Loliondo, for which a Presidential License was issued. . . . Mweka also had at least one hunting safari there in the '60s and their hunting was covered by a blanket Presidential Licence, I think."

The legal status of the reserve was later changed to that of a game controlled area (GCA) to allow for commercial trophy hunting before it was allowed in game reserves.

In 1992, Ali Hassan Mwinyi used his presidential prerogative to sell a ten-year concession to the Ortello Business Company (OBC), an obscure organization based in the United Arab Emirates (UAE) that was owned by Brigadier Mohamed Abdul Rahim Al Ali, a member of the UAE royal family. The deal was reportedly worth U.S.$1 million. Ortello was licensed in 1993 to carry out hunting activities and allocated hunting blocks in the Loliondo GCA. OBC management proceeded to divide the whole GCA into two hunting blocks covering virtually all of the 4,000 sq. km.

Allocation of hunting blocks to OBC was opposed by the Masai resident in the Loliondo GCA, who said their land had been allocated to the Arab company without their consent. Furthermore, although 90 percent of the Loliondo GCA consists of pastoral communities, the Masai were allocated just 17 percent of the controlled area—which they were also required to share with farmers. But their continuing resistance, including failed appeals to the courts, was futile.

Although this is a marquee example, in fact, allocation of areas to hunting companies has rarely taken into account the land rights of the communities resident in the GCAs. Never mind that this class of protected area has a provision that hunting on private land requires the consent of the landowner (titled or certified village lands qualify according to the GCA definition of private lands). In general, local communities have no role in the block allocation process.[15,16]

The result is that tourist hunting is conducted extensively on community lands without the permission or involvement of the landholder. The president still retains the right to grant concessions, for hunting, large-scale agriculture, biofuels, mining, hotels, and so on. He may delegate this authority to the commissioner of lands and settlement.

While no other safari outfitters have been granted permits in the Loliondo controlled area, some Masai villages have exercised their traditional land rights to lease land to ecotourism companies. This despite disputed rights of passage from OBC guards armed with AK-47s. (I can vouch for this because I too was accosted by OBC guards while leading a camping safari in 2001.)

Ortello (a.k.a. United Emirates Safaris) uses a maximum of four months of the six-month hunting season. But the scale of hunting cannot be compared to even the wealthiest hunting outfitters. Hundreds of members of Arab royalty and high-flying businessmen spend weeks in the Loliondo GCA each year participating in large-scale hunts operated like a military exercise. The hunters fly from the UAE in huge cargo and passenger planes, which land on an all-weather airstrip inside the OBC camp.

Frozen game meat shot on Loliondo hunts, as well as live mammals and birds destined for zoos and private collections, was regularly flown back to the UAE on the big cargo planes.

But why the Arabs have a special status is revealed in the other side of the story. In response to media reports supporting Loliondo Masai land rights, the minister for natural resources and tourism pointed out that the Masai, despite agreeing with Ortello not to do so, had, in 2006, started constructing new bomas (settlements), farming, and bringing in large numbers of cattle during the hunting season, at the instigation of local NGOs and some photography operators in the Loliondo area.

One large-scale hunt on August 13–15, 2001, was reported by members of the Masai Environmental and Resource Coalition (MERC).

Masai employees witnessed a hunting expedition with King Abdullah II of Jordan, accompanied by a very large entourage. Masai movement of the local people was severely curtailed around Oltigomi, Ololsira-Lukunya, and Ololosokwan areas. Numerous vehicles with radios, a helicopter, and two small planes patroled the entire area. King Abdullah's entourage used these helicopters, small planes, and vehicles to herd wildebeest and other large groups of wildlife toward the foot of the hills for easy hunting. For two days, MERC heard gunshots ring out almost continuously from morning until the late afternoon. The observers estimated that at least 60 animals were killed or wounded in the two-day hunting expedition.

Ortello annually paid $560,000 to the central government, $150,000 to the eight villages around the Loliondo Game Controlled Area, and $109,000 to the Ngorongoro District Council, as well as contributing vehicles, transceivers, and field gear to the Wildlife Division, financing construction of Waso Primary and Secondary Schools and six bore holes and cattle dips, purchasing two buses to enhance local transportation, and constructing various buildings and roads. No other district in Tanzania containing hunting areas receives anywhere near this level of funds for community development.

All things considered, it seems doubtful the UAE's tenure will be revoked anytime soon. It is somewhat reassuring to know that OBC's comings and goings are more closely supervised by the Game Division and the Immigration Department than in the freewheeling 1990s. The conflict over land rights between the Masai and the Arab hunting concession is ongoing, however, and the government still exercises the presidential authority to decide what is in the national interest.

According to the Village Land Act of 1999, all land in the Loliondo Division was classified as Village Land. But the GCA leased to OBC in 1992 cut a wide swath through the village land, including vital dry-season pastures adjacent to the national park.

Masai who moved their cattle into the GCA were harassed by OBC security forces frequently supported by the police. Whether OBC had the right to keep Masai and their cattle off their concession is debatable, as settlement, cultivation, and grazing as well as tourist hunting were permitted in GCAs.

So the conflict between the two sides remained unresolved—until the Wildlife Conservation Act of 2009, which prohibited cultivation and grazing in a GCA. Masai from many villages joined together in an unsuccessful court challenge. Those who had settled inside the GCA were ordered to leave. When they refused, a special police force, the Field Force Unit, managed by the district and regional authorities, forcibly evicted the residents of eight villages from the disputed area adjoining the national park. As widely reported in the regional and international press, the Field Force burned some 150 huts, grain stores, and maize fields, beat up or arrested anyone who resisted, and drove some fifty thousand cattle back into the Masai village lands in a time of severe drought. (The information above is from a report prepared by the Tanzania Natural Resource Forum, based on technical support and background research from Maliasili Initiatives, February 2011.)

The latest conflict between the government and the Loliondo Masai arose when the minister of natural resources and tourism announced on March 26, 2013, that 1,500 sq. km of the 4,000 sq. km Loliondo Game Controlled Area would be set aside as a wildlife corridor connecting the Ngorongoro Conservation Area with Serengeti National Park. He said the corridor was essential for protection of wildlife and water catchments. Local communities would not be allowed use of the land, and six communities presently in the GCA would have to be relocated. However, wildlife would not be protected from hunting, as OBC would have access to the full 1,500 sq. km corridor, right up to the boundary with the national park.

News of this apparent violation of the Masai's land rights led to an international outcry, spearheaded by NGOs like Cultural Survival whose mission is to protect the rights of indigenous peoples. Thirteen mostly Tanzanian pastoral and human rights NGOs issued a press statement on April 11, 2013 denouncing the government's action as a naked land grab for the sole benefit of the OBC.

On May 22, an international consortium of local communities, indigenous peoples, and NGOs, the Tanzanian National Resource Forum, "expressed concern to the Tanzanian Government that the proposed Loliondo Game Controlled Areas will threaten both conservation and local livelihood interests." The letter points out that prohibiting community residence or livelihood activities within the GCA would deny Masai cattle keepers access to traditional dry-season grazing grounds. An estimated twenty thousand people would be adversely affected.

As all forms of hunting are prohibited in Tanzania's national parks, I find it curious that OBC will be allowed to shoot animals in the wildlife corridor, protection of which will involve SNP and NCA wardens and rangers. I was also surprised to find that Lobo Lodge, the first lodge erected inside the park, has a connection to the OBC. A brief stop in July 2012 while waiting for the airplane that would fly my party back to Arusha following a walking safari in the adjacent park wilderness area, showed a facility quite transformed from the orginal. Two years earlier, I had found Lobo in a state of disrepair, with poor service. In the meantime it had undergone a makeover and now presented a decidedly Near Eastern decor. The manager proudly showed us the VIP suite, which was decorated in a manner that should make UAE sheiks and princes feel at home. In fact, he said it was regularly used by guests of the OBC. He pointed out a well-used road that he said connected the OBC hunting concession with Lobo Lodge.

How Secure Is Serengeti National Park?

Hilborn and colleagues[17] wrote in 2008, "In comparison with other natural areas, the Serengeti ecosystem remains relatively undisturbed. Fences have not been built, bore holes have not been drilled to provide supplemental water, major predators have not been eliminated, antipoaching activities have restricted the level of harvest, and the geographic integrity of the migration is largely intact. In comparison to Yellowstone or Kruger national parks the Serengeti ecosystem is reasonably pristine" (436). However, the Serengeti NP General Management Plan 2006–2016 mandates major changes in the present hands-off policy. Stay tuned.

Poaching in the Serengeti was rampant from the mid-1970s to the late 1980s, during the time when Tanzania was so poor it couldn't pay rangers or provide fuel for patrols. Rhinos, estimated to number over 700 in the mid-1970s were shot out in about three years. Only a dozen or so survived in the park by 1977. The elephant population of 2,500–3,000 was reduced to about 500 in Tanzania's part of the ecosystem. Many more found refuge in the Masai Mara, where the presence of tourists complicated poaching activities. There were very few tourists in Serengeti NP after Tanzania closed the border in 1977 to stop Kenya tour operators from dominating the tourist trade. Elephant poaching finally dropped off in 1989, when the CITES ban on ivory trading was imposed and the Tanzania government authorized Operation Uhai (Life) permitting shooting of poachers on sight. Before that, poaching in the Western Corridor and on the Wakuria tribal land in the Kogatende area reduced the buffalo population from 60,000 to 20,000. And one iconic predator had been lost: wild dogs infected with rabies and canine distemper from neighboring domestic dogs disappeared from the park in the 1970s.

Pressure on Serengeti NP from the Growing Human Population

The population of mainland Tanzania almost tripled from 1967 to 2002, from 11.96 million to 34.6 million. At the present growth rate of 2.3 percent, the country is expected to have about 56 million people in the year 2020.[12] A Tanzanian biology professor, M. Songorwa,[18] has been a "voice in the wilderness," warning of a looming crisis in wildlife (and other natural resources) conservation that will become catastrophic if appropriate measures are not taken now.* Tanzanian wildlife managers, conservationists, and policy makers are "aiming at and 'shooting' the wrong target(s)" by focusing on symptoms instead of the root cause of conservation problems—demographic factors.

Songorwa calls on the developed nations, most of whom, by destroying their own natural resources, now have the necessary money and expertise to help save Tanzania's natural treasures.

Pressure Points on the West Side of Serengeti NP

According to A. B. Estes et al.:[19]

> The growth of human populations around protected areas accelerates land conversion and isolation, negatively impacting biodiversity and ecosystem function, and can be exacerbated by immigration. It is often assumed that immigration around protected areas is driven by expected economic benefits, but in many cases people may be pushed from their areas of origin toward protected areas. . . . We found that conversion of natural habitats to agriculture was greatest closer to the park (up to 2.3% per year), coinciding with the highest rates of human population growth (3.5% per year). Agricultural conversion and population growth were greatest where there was less existing agriculture, and population density was lowest (255).

---

* Professor Songorwa, recently appointed director of the Wildlife Division, is now in a favorable position to gain the needed international financial support and wildlife expertise.

Tanzania is hardly the only country failing to take demographics into account. An increase of 2–4 billion people in the world population over the next thirty years is going to have a devastating effect on most other countries' carrying capacity, not to mention nature conservation. Yet you can hardly find a politician who will even discuss the population explosion, especially in the U.S. where limiting population growth is the real third rail of politics. Some presidential candidates would prohibit all efforts at family planning, proud to have set an example by producing up to seven offspring of their own.

That population growth in the western Serengeti is driven by people being pushed from densely settled areas in search of resources is shown by the fact that "about three-quarters of households directly adjacent to the park in the northwest (75.1%) and southwest (71%) are below the poverty line,[20] as compared to 42% and 46%, respectively, of the larger region to which these communities belong[21] and 39% of rural populations in Tanzania as a whole[22]" (261).[19]

The boundary between settled agricultural land and the savanna vegetation that dominates most of the park is, in many places, so abrupt that it is easily visible on satellite imagery (map 12.1). This indicates a breaking of the ecological link between the park ecosystem and its western surroundings. On the eastern side, the park is still linked to its surroundings in the adjacent Kenyan and Tanzanian Masailand.

"Protected areas will continue to be engulfed by advancing agricultural frontiers," Estes et al. predict,[19] "as people are forced to leave their areas of origin to access arable land.[23] If protected areas are to persist as havens of biodiversity, then early detection and analysis of the factors contributing to encroachment are needed to facilitate land use planning to mitigate the consequent environmental impacts" (256).

Bushmeat Dynamics

Bushmeat is the primary source of meat-based protein for the majority of households living in the western Serengeti; 95 percent of households considered bushmeat one of their most important protein sources.[24] Cross-sectional surveys of hunting practices and consumption patterns have found that bushmeat was consumed regularly by 45 to 60 percent of households in the northwestern Serengeti,[11,24,25] with migratory ungu-

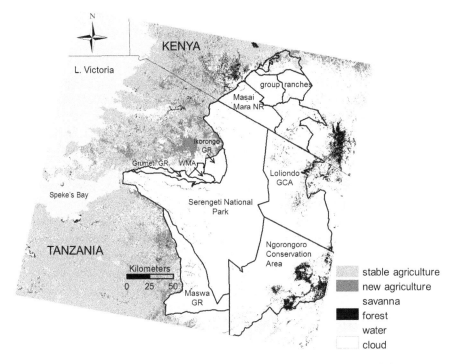

MAP 12.1. This map shows that agricultural development had already occupied the whole region west of the Serengeti protected areas in 2002. From A. Estes et al. 2012.

lates (wildebeest and zebra) accounting for an estimated two-thirds of total offtake.[26] Migrant herbivores are especially at risk as predictable seasonal movement patterns make them easy targets of poachers' snares.

Bushmeat Consumption Research of Dennis Rentsch. Rentsch[27] estimates an annual offtake of approximately 77,000 to 129,000 wildebeest, or approximately 6 to 9 percent of the current population. Previous estimates using an ecological model of wildebeest dynamics based on long-term data sets of wildebeest population trends and rainfall have been much lower: no more than 40,000 per year or 80,000 per year.[28] Yet evidence based on surveys of bushmeat consumption and poachers' kills demonstrates that annual offtake is greater than 80,000[11,24,25]. Furthermore, recent population estimates, including the latest (2009) aerial survey, fail to show a downward trend in wildebeest numbers, probably because of the unrealistic assumptions of the demographic model.

The demographic model was based on estimates of wildebeest recruitment data from 1992–93, when the population suffered a severe drought (see chapter 11) and the human population in western Serengeti was approximately 370,000, compared to nearly 570,000 today.[22] More important, the ecological model of wildebeest population regulation assumes that there is an equal sex ratio and that equal numbers of males and females are poached.[28,29]

In fact, males are far more likely than females to be caught. Bulls are generally in the vanguard of moving aggregations and therefore more likely to pass through snare lines, and they typically migrate through human-inhabited areas ahead of female herds. Bachelor males made peripheral to a spread of females by territorial bulls spend more time in wooded areas infested with snares, and territorial bulls prefer to stand next to trees.[30,31] Preliminary evidence of a male-biased hunting offtake of 3.5:1 would allow more than 100,000 wildebeest to be removed by hunters each year.[25,33–35] If cows are still cranking out 400,000 to 500,000 calves per annum, that would explain the consistency of survey results.

If the current rate of human population growth along the west and southwest side of the Serengeti (approximately 3.2 percent per annum) is sustained, the population will double in twenty-three years, from 600,000 in 2007 to 1.1 million in 2026. Because wildebeest make up the largest biomass of mammals and their migration is the defining attribute of the ecosystem, any existential threat to wildebeest is a fundamental conservation issue.

Offtake of wildebeest cannot be sustainable while nearby human populations continue to grow at an exponential rate and demand for bushmeat per capita remains at current levels. New strategies to address bushmeat hunting are therefore urgently needed if the wildebeest population is to hold its own. The fact that demand reaches beyond the local communities and beyond the ecosystem must also be considered. Growing human population pressure means that demand for land for grazing livestock and fertile agricultural land is increasing while the fish supply from nearby Lake Victoria is decreasing. Despite recent heavy investment in antipoaching and efforts to improve detection rates for poachers,[2,17,36] demand for bushmeat remains high. As local human populations continue to expand, the costs of abating illegal hunting through community outreach initiatives as well as enforcement tactics will also climb. Demand for bushmeat is unlikely to be reduced by antipoaching alone as households continue to search for cost-effective protein.[27]

Current and Proposed Development in Serengeti National Park (a.k.a. SENAPA)

According to the SENAPA General Management Plan 2006–2016, "an optimal balance between protection and use of the Park has been established" in consultation with all major stakeholders. The most important points, taken from the executive summary, are summarized here.

A new zoning scheme divides the park into three management zones, replacing eight in the previous management plan. The zoning aims to provide a framework for achieving and reconciling the twin management needs of protecting the natural qualities and environment of the park and regulating and promoting visitor use.

> *The High Use Zone* covers 23 percent of SENAPA and is centered on the Seronera Valley. The General Management Plan (GMP) stipulates that any increase in bed capacity will be matched by an equivalent expansion in services and roads.

> *The Low Use Zone* covers 42 percent of SENAPA and is contained in five blocks distributed evenly throughout the park. The GMP stipulates that only permanent and nonpermanent tented camps will be permitted in this zone, with visitor activities focused on game viewing by vehicles. Up to 188 additional beds are proposed, including up to four new permanent tented camps around Wogakuria and south of the Grumeti River at Musabi and thirteen new "Special Campsites" in more remote areas. This is the only zone where off-road driving may be permitted in designated areas, which will be identified annually by SENAPA management.

> *The Wilderness Zone* covers 35 percent of SENAPA and incorporates the hilly areas of the park, where the very limited road network and lack of any facilities already restricts access. The only visitor activity permitted is walking safaris; game viewing by vehicles is prohibited. This zone contains no permanent structures, apart from a limited number of access roads to wilderness campsites on the zone edges for use by SENAPA management and the support teams of walking safari operators.

Park management is divided into four departments:

1. Ecosystem Management (Ecology and Protection Departments)
2. Tourism Management (Tourism Department)

3. Community Outreach (Outreach Department)
4. Park Operations (Protection, Administration and Stores/Works Departments)

"The Tourism Management Programme aims to provide an outstanding experience for both local and international visitors, optimal economic benefits to the nation, TANAPA, private sector partners and local communities, and minimal impacts on the Park's resource values. A high priority for this GMP will be the enhancement of visitor access and use while at the same time minimising disturbance to key habitats and wildlife. . . . Another key strategy to achieving a world class and environmentally responsible tourism experience in the Serengeti will be to upgrade and expand the existing SENAPA visitor facilities. These improvements will be done in close consultation with the tourism industry, with priorities being to upgrade water supply and sanitation throughout the Park and to provide accommodation appropriate to the needs and budget of local visitors, an important step in encouraging more Tanzanians to visit the Serengeti.

"Park operations: The GMP problem analysis identified poaching as the priority management issue to be addressed by this Programme. In response, SENAPA management will investigate and pilot new anti-poaching techniques; re-equip the Protection Department with the necessary modern equipment; build ranger-local community cooperation and anti-poaching reward schemes; and provide training to Village Game Scouts to protect conservation areas in the SENAPA buffer areas."

Paradise Lost?

Far be it from me to second-guess the GMP, which the most experienced and dedicated conservationists, scientists, and protected-area managers believe is the best hope for preserving the Serengeti ecosystem. Implementation will undoubtedly improve the present system in many ways. The losses that I mention are rooted in nostalgia for the way things were in the good old days, when there were many fewer people and many fewer restrictions.

For instance, barely a decade ago driving off-road was permitted everywhere outside a 12-mile radius from Seronera and in the Triangle between the two roads to Ndutu. That area was made off-limits in the 1970s, when wild dogs denned there. (Not long afterward, wild dogs went extinct in the plains, but the restriction was never lifted.) Now driving off-road except in designated areas within the Low Use Zone can be severely punished: park rangers are so quick to apprehend offenders that tour drivers are reluctant to leave the road even where permitted for fear of hefty fines or being banned outright from the park.

The speed limit of 45 kph (30 mph) throughout the park is rarely observed except on game drives. Consequently, much of the road between Naabi and Seronera becomes so "washboard" that it is necessary to exceed the speed limit to make the drive tolerable.

The projected development of tourist facilities will greatly increase the human footprint in the ecosystem. Hotels and lodges have the greatest environmental impacts. When my family lived in the Serengeti in the late 1970s, there were just two lodges inside the park, at Seronera and Lobo. When the economy picked up in the 1980s and the Kenya border was reopened, the need for new accommodations was recognized. But park trustees agreed with researchers and conservationists that hotel and lodge construction should be kept outside Serengeti NP. That agreement was breached when Sopa was allowed to build a lodge in hills at the southern end of the park, allegedly brokered by top government officials. TANAPA's director general resigned in protest. The next big permitted development was Serengeti Serena Lodge, constructed off the road from Seronera to Kirawira.

As of this writing there are a total of six lodges in the park: Seronera and Lobo Wildlife Lodges, Sopa Lodge, Serena Lodge, Mbalageti Lodge, and Bilila Lodge Kempinski (now the Four Seasons Serengeti) (fig. 12.2). Then there is an array of permanent camps throughout the park, including Kusini, Dunia Camp, Kati Kati Tented Camp, Mbuzi Mawe, Kirawira Tented Camp, Grumeti River Camp, Sayari, Lamai Serengeti, Serengeti Bushtops, and Kuria Hills. In addition, there are an increasing number of mobile tented camps that operate year-round in the park but move twice yearly to be closer to the migration.

With the GMP plan calling for an additional 188 beds and up to four new permanent tented camps in the Low Use Zone, dare one ask how all these developments are compatible with "minimal impacts on the Park's resource values."

All I can say is, thank God for the Wilderness Zone! I led a camping-walking safari there in 2010 and another in June 2012. It was like seeing the African savanna in its primeval state. No other people! Hills and river valleys teeming with all the usual big and small game too distant from settlements and poachers to fear people on foot. It was as if my dream of coming to Africa a century ago had come true.

But expensive! Serengeti NP fees, payable in foreign currency, have been going up year by year, purportedly to maximize income while keeping down the number of visitors. The daily admission fee of $60 for noncitizens is just the beginning. Camping fees per person

Bilila Lodge, the latest addition (2009), advertised itself as "the Largest Lodge Ever Built in the Serengeti National Park, 74 rooms in an untouched part of the Park where the hotel will have its own private game drives." Among many other amenities, "all the guest rooms have luxury ensuite bathrooms, multimedia DVD players, over 50 satellite TV and radio channels, coffee/tea making facilities and a teak deck for relaxing while watching game (personal telescope included); suites have their own private plunge pool." Guests could wonder, is going on a game drive really worth the effort?

Indeed, this was an almost untouched (except a Serena tented camp) part of the park, near the wonderful Mbuzi Mawe (Rock Goats = klipspringer) kopjes you pass en route to Lobo. To my mind, allowing this intrusion was even more inexcusable than permitting Sopa Lodge, and begs the question, who was behind it? (Weekend stays by Tanzania's president led to some speculation.)

My concerns were shared by the UNESCO World Heritage Committee. At its thirtieth session the World Heritage Committee, together with IUCN, raised questions about the lodge development.[32] These included its impact on surface water and groundwater quality, compatibility with the General Management Plan and Tanzanian National Park policies, sustainability of visitor numbers, and prevention of overcrowding, particularly in sensitive areas.

It is now the Four Seasons Safari Lodge, and its new management is creating a Discovery Center intended to raise guests' appreciation of the Serengeti ecosystem. Serengeti scientists will be encouraged to give talks and involve guests in aspects of their studies. The idea of supporting an in-house student doing research for a MSc or PhD has been raised, and a resident naturalist may be brought on board. According to Oli Dreike (pers. comm. October 27, 2012), who is designing the Discovery Center, "There seems to be a real commitment to make this work and create something like no other safari lodge in Africa." If these efforts bear fruit, they will help mitigate the ecological impact of this major development.

vary depending on location, topped by a daily rate for the Wilderness Zone.

TANAPA even refuses to waive admission fees for former researchers who have made major contributions to understanding the ecosystem. This seems to reflect rooted suspicions that foreign researchers exploit Tanzania's treasures for their own benefit. Some individuals have, indeed, carried out projects of no immediate or direct benefit to the park, and a few have made little effort to share their results with

FIGURE 12.2. Photo of Bilila Lodge, now owned by Four Seasons. Despite the size of this development, wildlife are reclaiming their space. Photo by Oli Dreike

TAWIRI or Serengeti NP managers, but in general researchers try hard to work with park management, including helping to report poaching and tourist infractions to park authorities. Nonetheless, the management's restriction of once-common research activities for fear that they may be frowned upon by tourists seems to treat research as a secondary rather than a primary function and responsibility of national parks. Instead, management could choose to view research as an asset and work together with the research community to raise the profile of research and its importance to a generally curious and receptive tourist public.

Access to Lake Victoria for Serengeti Wildlife

TANAPA plans to expand the Serengeti National Park to Lake Victoria by annexing the Speke Gulf Game Controlled Area (www.tanzania-parks.com). This salient of Lake Victoria is within 3 km of the park boundary and surely used to be an important permanent water source for Serengeti wildlife, including the wildebeest and zebra migrations (see map 12.1). But the area is now so heavily settled, including a lodge and substantial vacation homes, that wildlife would have to thread the needle to reach water. Yet, given the imminent danger that the Mara River could cease to flow year-round (explained later), the lake is the only other guaranteed permanent source of water and may well be critical to the future of the ecosystem. The best that could be hoped for is a fenced corridor from the park to the lake, and even this may soon be impossible because of increasing settlement. Meanwhile, the lake's water level is falling, altering the shoreline and increasing the distance from the present park boundary. (See the Uganda section below.)

Proposed Highway and Railroad across the Serengeti

Plans to construct a 33 mi. (53 km) roadway across the northern Serengeti NP to link a proposed paved highway from Lake Victoria to the port of Tanga on the Indian Ocean were announced during the 2005 national election won by President Jakaya Kikwete, who promised to begin construction in 2012 (map 12.2). In the run-up to the 2010 national election, in which Kikwete faced his most serious challenge to leadership of the ruling CCM party, he renewed his pledge to construct the highway in order to assist development and market access for the isolated populations on either side of the park. In response to increasing foreign demand, it would also be used for transporting minerals and oil, from Central Africa to the coast. By 2010, the roads department had already staked out the proposed route with flags.

Serengeti scientists, conservationists, and the UN were quick to point out that the road through the park would bisect the ecosystem and block the migration. The Frankfurt Zoological Society warned in a June 15, 2010, statement that if access to critical dry season resources in the Masai Mara were cut off, "the population would likely decline from 1.3 million animals to about 200,000 (meaning a collapse to far less than a quarter of its current population and most likely the end of the great migration)." A letter published in *Nature* signed by twenty-seven scientists (including R. D. and A. B. Estes) was titled, "Road Will Ruin Serengeti."[37] The protests spread worldwide through the news media and over the Internet. But Kikwete refused even to consider a better alternate route south of the Serengeti, even though it would serve far more communities than the northern road.

Unexpected levels of international condemnation resulted, including promises to boycott Tanzania as a tourist destination. Potential investors, international aid agencies, and NGOs became unwilling to risk criticism by supporting the proposed highway. On June 23, 2011, the Tanzanian Ministry of Natural Resources and Tourism backed down, and there was a collective sigh of relief among all those concerned. The alternative route well south of the park had won.

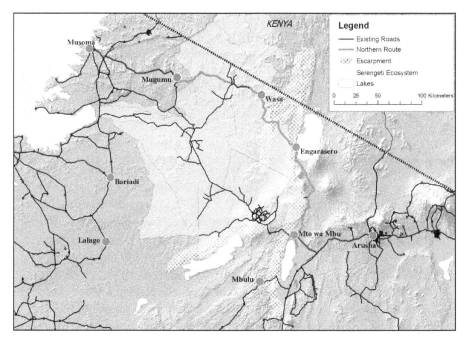

MAP 12.2. The proposed highway across Serengeti NP and the alternate southern route finally(?) accepted. One of the FZS Highway maps by Grant Hopcraft, posted on-line.

But the celebration turned out to be premature. Even more alarming, Tanzania and Uganda agreed in April 2011 to build a railway line connecting Kampala with the coast, which would follow the same route as the canceled road. Capable of transporting oil estimated to be worth $2 billion a year for the next twenty years, the proposed railroad, according to President Yoweri Museveni, was to be "the lifeline of the Uganda of his dreams."

On December 28, 2011, the *Kampala Daily Monitor* reported that the Chinese intended to complete a feasibility study within six months so that construction could begin in April 2012 and be completed in 2015. Once again, there was a global outcry. This time lawsuits were filed against the proposal with the East African Community. And once again a government ministry changed its mind. On January 8, 2012, Transportation Minister Omar Nundu told the (Tanzania) *Guardian,* "Rest assured that the railway line will be constructed 100km south of the Serengeti national park's sprawling expanse."

These threats to the future of the Serengeti ecosystem have been laid to rest, hopefully for good. But they serve as a warning that protected areas will remain at risk as pressures to modernize and develop resources and industry to support ever-growing populations increase. (And as detailed later, it remains uncertain that Tanzania has really given up all plans for a roadway or railway across the Serengeti.)

Kenya

A British protectorate from 1895, then a British colony in 1920, Kenya became an independent republic in 1963 under Jomo Kenyatta, leader of the Kikuyu, the dominant tribe. His political party, the Kenya African National Union (Kanu), became a de facto one-party state after absorbing the party of Vice President Oginga Odinga, leader of the Luo, Kenya's third-largest ethnic group, when he attempted to form an opposition party in 1966. Following Kenyatta's death in 1978, Vice President Daniel arap Moi, head of a minor tribe, the Kalenjin, was elected head of Kanu and designated the sole nominee for the presidential election. In 1982, the National Assembly amended the constitution to make Kenya officially a one-party state.

Street protests and international donor pressure caused parliament to repeal the one-party section of the constitution in 1991. A year later independent Kenya's first multiparty elections were held. Arap Moi and Kanu managed to stay in power until 2002, when a coalition of opposition parties won the presidency and parliament. Mwai Kibaki, Kenya's third president, who had held senior posts in both the Kenyatta and Moi governments, was also a Kikuyu.

Subsequently, the Kenya political system has been plagued by internal conflict, corruption scandals, charges of rigged elections, and, in 2007–8, ethnic violence.

Two former Kibaki allies, Raila Odinga and Kalonzo Musyoka, formed the coalition Party of National Unity (PNU) and ran against Kibaki in 2007, each supported by a different faction. In the ensuing general elections, the presidential vote was widely regarded as rigged by PNU, with glaring irregularities in the vote tabulation process as well as turnout in excess of 100 percent in some constituencies. When the chairman of the Electoral Commission declared Kibaki the winner, violence erupted in different parts of Kenya as supporters of Odinga and of Kibaki clashed with police and each other. By the time it ended, 1,300 Kenyans were dead and 500,000 were displaced as neighbors of tribes that had lived together peacefully suddenly became enemies. In 2010, five high-ranking government officials were charged with crimes against humanity for their roles in the 2007–8 postelection violence—including the just elected president, Uhuru Kenyatta, son of Jomo.

> Needless to say, this explosion of ethnic violence had an immediate though short-lived effect on tourism in East Africa (while boosting the South African market). I lost my summer job as resident naturalist at Governor's Camp in the Masai Mara, as there were too few guests to justify my continued employment.

In 2008, brought to the table by the UN secretary general, President Kibaki and his rivals agreed to a power-sharing arrangement that included the position of prime minister, to be filled by Odinga, two deputy prime ministers, and cabinet posts based on the parties' proportional representation in parliament. Parliament ratified the agreement with a new constitution. The forty-two-member cabinet became the largest in Kenya's history—also the most highly paid, allegedly with higher salaries than members of the U.S. House of Representatives.

A new constitution, overwhelmingly approved in a 2010 referendum, retains Kenya's presidential system but introduces additional checks and balances on executive power and greater devolution of power to the subnational level. The 2013 national elections were the first conducted under the new constitution.

Land Tenure

Freehold land tenure was introduced by the colonial administration to guarantee property rights of white immigrants who otherwise were reluctant to risk settling in Kenya. These and other colonial policies were continued after independence, when Kenya became a member of the Commonwealth. Landownership is facilitated by Kenya's predominantly "free market" economic policy. Government-owned parastatals, including the Kenya Electricity Generating Company (KenGen), Telkom Kenya, and Kenya Re-Insurance, were privatized in 2008. Kenya is one of the most capitalistic nations in Africa.

However, U.S. State Department Bureau of African Affairs Background Note: Kenya (January 11, 2012) indicates there is considerable room for improvement.

> Tourism rebounded from the drop experienced in 2008, bringing in $807 million in 2009, and in 2010, the Kenyan Ministry of Tourism recorded nearly 1.1 million tourists—an all-time high. Nevertheless, Kenya faces profound environmental challenges brought on by high population growth (about 2.5 percent), deforestation, shifting climate patterns, and the overgrazing of cattle in marginal areas in the north and west of the country. Significant portions of the population will continue to require emergency food assistance in the coming years, especially the many thousands displaced during the 2007 bloodbath who still live in refugee camps.
>
> Accelerating growth to achieve Kenya's potential and reduce the poverty that afflicts about 46 percent of its population will require continued deregulation of business, improved delivery of government services, addressing structural reforms, massive investment in new infrastructure (especially roads), reduction of chronic insecurity caused by crime, and improved economic governance generally.

Kenya Wildlife Service Responsibilities

To say that Kenya's wildlife protection performance has had its ups and downs is putting it mildly. In colonial times, the Kenya Game Department and National Parks were separate organizations that continued to be headed for a good decade after independence by Maj. Bruce Kinloch and Col. Mervyn Cowie, respectively, two of the most dedicated conservationists I have known. Following their handover of leadership, corruption in the Kenya government soon spread to the Game Department. The Wildlife (Conservation and Management) Act of 1976 amalgamated the Game Department and the Kenya National Parks to form a single agency, the Kenya Wildlife Service (KWS), under the Kenya Ministry of Tourism and Wildlife.

Charged with conserving and managing Kenya's wildlife resources, KWS is custodian of Kenya's twenty-six national parks and thirty national reserves, as well as wildlife on otherwise unprotected private and trust lands, where more than 70 percent of Kenya's wild animals live. It also is responsible for preserving ecosystems and biodiversity and, according to the KWS website, "ensuring that the nation's wildlife-related resources remain in optimum condition for the multiple activities the government and local people demand of them."

The most noteworthy subsequent history of the KWS occurred under the leadership of two Kenya citizens of English ancestry: Richard

Leakey, archaeologist son of Louis and Mary Leakey; and David Western, a leading ecologist famous for his research and conservation efforts in the Amboseli ecosystem. Leakey used his fame and prominent political status (including his own political party) to raise millions of dollars for KWS. The old colonial HQ at Nairobi NP were replaced by multiple modern office buildings, a museum, a restaurant, a zoo, and an animal orphanage.

Leakey's primary mission was protection and management of the national parks. Western took the opposite tack when he became director. He made KWS responsible not only for the national parks, but for all the wildlife on the 70 percent of unprotected land outside the parks. This huge increase in the responsibilities of park wardens and rangers soon led to reduced management and protection of the parks and reserves. Nor did efforts to protect wildlife on the outside do much if anything to reduce rampant commercial poaching, which provided much of the meat sold in Nairobi and other population centers.

Partly in the hope of controlling poaching, Kenya banned hunting in 1978. The political and economic influence of animal rights NGOs has kept the ban in force ever since, despite the economic benefits landowners formerly gained from exploiting the wildlife on their property. Absent incentives for landowners to tolerate crop damage and other wildlife-related problems, wildlife outside protected areas has declined by an estimated 70 percent.[38] At present, most landowners feel that under current law and management wildlife is a liability imposed on them. They are desperate for relief.

Recent changes in KWS's approach to wildlife conservation provide a glimmer of hope. Instead of maintaining expensive control operations to eliminate problem animals as the need arose, KWS now prefers a strategy that encourages integration of wildlife management objectives with those of the landowners. The aim is to establish sustainable wildlife utilization as a viable land-use option in areas outside national parks and reserves. Toward this end, KWS has established a pilot extension service, the Community Wildlife Service (CWS), which encourages landowners in selected districts to accept wildlife on their land in return for benefits, such as sharing in revenues from consumptive utilization of wildlife and from tourism. Elsewhere outside protected areas, KWS intends to encourage conservation by assisting landowners and local people living near parks in obtaining various tangible benefits from their tolerance of wildlife, including sharing of park revenue and promotion of wildlife-based economic activities, principally tourism and certain forms of wildlife utilization.

An even more radical departure from the previous no-hunting policy can be found on the KWS website.

> Most individual ranchers and ranching companies want the ban on hunting lifted to allow sport hunting. Sport hunting, they say, earns landowners maximum yields from wildlife and is therefore preferable to cropping. Many group ranchers are most flexible in their views on utilization. They are ready for incremental development, starting with bird shooting and commercial cropping of plains game. Kenyan hoteliers and tour operators generally are averse to sport hunting and insist that if it is done, it must be kept cryptic.
>
> For people participating in the CWS project, the major point of concern is the illegal nature of its pilot use-right scheme. The only provision for hunting in the law is the KWS Director's Special Authorization to hunt, which is intended for application in special and limited circumstances such as research and does not provide for commercial operations. Bans on trade in wildlife also are still in force in Kenya. (Although such activities are not illegal, the country lacks policy provisions covering private game reserves, orphanages, game farms, domestication, etc., as well.)
>
> Thus, individual ranchers in Laikipia and Machakos would like to be allowed the same rights to bird shooting, game cropping, game farming, and ecotourism activities as presently enjoyed by group ranchers in Samburu and Kajiado, but they feel that they can make no major investments to benefit from such activities until the legal status of hunting and trade in wildlife products is reversed.

Protection of Kenya's Part of the Serengeti Ecosystem

If only these proposed changes in wildlife policy could be more widely applied on the Kenyan side of the ecosystem! Instead, private landownership prompted landowners to respond to market opportunities for mechanized agriculture at the expense of wildlife habitats.[39,40] The value of this option was fifteen times greater than the alternative use for wildlife-based tourism and limited agriculture and livestock.[41]

Over 50,000 ha of rangelands, primarily on group ranches, have been converted into large-scale mechanized wheat farms.[42] This resulted in serious destruction of the core breeding and calving grounds and the crash of the migratory Mara wildebeest population (see chapter 4). Resident nonmigratory wildlife populations, including giraffe, topi, buffalo, and warthog, dropped by 73 to 88 percent, while waterbuck, Thomson's gazelle, Grant's gazelle, kongoni, and eland decreased by about 60 percent.[40] In the Masai Mara Game Reserve key ungulate populations have also decreased significantly largely due to poaching pressure.[43]

The evident change in KWS's mind-set harks back to the days when game ranching was permitted, notably by the Hopcraft Ranch in Athi River, which supplied game meat to Nairobi's famous Carnivore Restaurant. The logical next step would be to follow the path of Zimbabwe, Botswana, South Africa, and Namibia, where landowners own the wildlife.

It all started in Zimbabwe in 1975, when the Parks and Wild Life Act conferred to landholders custodianship of wildlife on their land. This was followed in 1982 by CAMPFIRE (Communal Area Management Programme for Indigenous Resources), which gave rural communities access to, control over, and responsibility for natural resources on their land. Virtually with the stroke of a pen, wildlife became valuable, for sport-hunting, game ranching, and tourism. Indeed, more valuable than livestock on many landholdings, leading to the establishment of conservancies, combining of neighboring properties into wildlife reserves, hunting blocks, etc. Private ownership led to the widespread reintroduction of game in areas where it had disappeared and even into areas outside the former range. That a prime sable bull may sell for $12,000 at a South African wildlife auction gives one a good idea of just how valuable wildlife ownership can be.

Because the rights of private property are fundamental in capitalistic Kenya, it is at least conceivable that wildlife ownership could be granted to landowners. Though possibly transformative, it would doubtless be politically impossible at present. It is no more likely to happen in Tanzania or Uganda than in the United States, where ownership and management of wildlife is vested in state fish and game departments.

The Kenyan government introduced group ranches in 1968 to address the problem of overgrazing and land degradation occurring in the pastoral arid and semiarid areas by converting communal to group tenure. This was expected to secure the Masai's traditional grazing grounds against alienation by non-Masai while encouraging nomadic pastoralists to confine their livestock within ranch boundaries and reduce their livestock numbers. The ranches were registered in the names of group representatives (3–10 members) elected by the members of the group.[44] They ranged in size up to 100,000–200,000 acres (40,468–80,937 ha) in the drier, less settled areas to only one-tenth that size in more populated areas with higher agricultural potential.

The Masai did not reduce the number of livestock they owned, nor did they restrict their livestock within the ranch boundaries, but to a large extent continued exploiting group ranch land along traditional lines.[45-47]

Group ranch subdivision into individual plots began in the 1980s with government support. Actually, privatization had been going on from the very beginning, as "those with seniority, influence, business acumen and or education" acquired individual portions of land within the group structures.[48] This led ordinary members to think about subdivision as a means of protecting their shares against the eroding integrity of the group domain.

Then lo and behold, the president of the republic proclaimed that all Kenyans had the right to own their own land and that group ranches should be subdivided among their members and abolished. That did it! Never mind that his interference was extrajudicial and outside the judicial system of governance and decision making in the ranches themselves.[49] The president then outdid himself by degazetting Amboseli NP and handing it over to the Masai district council.

Further fragmentation of plots and, in many cases, transfers of ownership to non-Masai followed subdivision. Subsequently, average plot size decreased, while the number of fenced properties and levels of cultivation increased. (Subdivision abutting the Mara Reserve also intensified competition for commercial campsites with convenient access, substantially increasing the number of visitors and creating still more congestion and disturbance in the reserve, not to mention the loss of wildlife habitat in the former ranches.) Due to subdivision, the Masai are gradually losing their best land and are being pushed into the drier areas. It threatens continued extensive nomadic livestock production by decreasing mobility and carrying capacity, increasing the potential for land degradation and crop failures, and interfering with traditional wildlife and pastoral migration patterns.[48] Pastoral peoples living in the Mara ecosystem have less livestock per person today than they did twenty years ago, and about half survive today on an income of less than $1 per day per person. If these trends continue, it is probable that the Mara will support very few wildlife and poorer pastoral peoples twenty years from now.

What If the Mara River Dries Up?

Poor stewardship of Kenya's natural resources and relentless pursuit of economic development have reached the point where the Kenya part of the ecosystem is at risk. Changes in the Mara River, of critical impor-

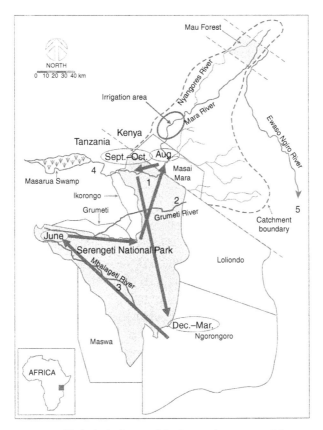

MAP 12.3. Hydrological map of the Serengeti ecosystem (Mnaya et al. 2011). Note locations of the Mau Forest catchment area of the Mara River and the major irrigation projects that draw water from the river .

tance to the entire ecosystem, are of particular concern. Pollution by pesticides, sewage, and increasing phosphate concentrations diminish water quality and threaten the survival of the Mara's diverse fish fauna, including species that disappeared from Lake Victoria following the introduction of Nile perch by the colonial administration.

In the worst case, the Mara River could stop flowing in the dry season. A main cause of its diminishing flow is the deforestation of its headwaters in the Mau Forest, which also contains crucial water catchments for some of Kenya's other large rivers. Although the Mau Forest has been officially protected since 1903, deforestation for settlement and agriculture (as revealed by satellite imagery) reduced the area covered by forest to 752 sq. km by 1973, 650 sq. km in 1985, and 493 sq. km in 2000. Following up on its privatization initiative, the government even proposed degazettement of the Mau Forest Reserve. By the time this policy was overturned, 89 percent of the forest was gone. Meanwhile, "important" people had taken over much of the land, making eviction of legal and illegal settlers that much more politically difficult. In addition to the Mara headwaters, the forests of the Mau Escarpment are crucial water catchments for some of the largest rivers in Kenya that feed such lakes as Nakuru, Bogoria, and Natron (map 12.3).

The other main cause of decreasing flow is pumping river water to irrigate 520 ha of mechanized farms in the Loita Plains, which has already reduced dry weather flow by up to 25 percent. Dry-season thunderstorms over the headwaters of the Mara and its tributaries seem to be more frequent nowadays and, lacking sufficient forest cover to absorb the downpour, cause flash floods that cut into the Mara's banks and whittle away the riverine forest.

As irrigation projects are of major economic importance, pumping may increase markedly in the future, thanks to the Kenya Water Act (2002), which allows abstractions of up to 70 percent of the total flow, leaving only 30 percent of the water for the river.

The drying up during a severe drought of the region's one perennial stream would cause the wildebeest and zebra populations to crash.[50] Therefore, maintaining the flow of the Mara River is critical to the whole ecosystem. That this may be so was demonstrated by studies of the Grumeti and Mbalageti, the other two major rivers, which are normally reduced to series of stagnant pools during the dry season. Most of these pools dry up during a drought year. From 1960 to 2000, this occurred six times, that is, once every seven years or so. The ecohydrological model predicts that within about two weeks of either river ceasing to flow, wildlife would start dying at a rate of 30 percent per week.[51,52]

The developers of the Serengeti ecohydrology model proposed a transboundary Mara River management plan, which would take into account the cost-benefit analysis (for Kenya and Tanzania) of deforestation, irrigation, and the proposed Ewaso Ngiro (South) Hydropower Project.[50] The idea was to request the Government of Kenya to act in the spirit of the East African Community and Regional Cooperation and not take unilateral decisions and go ahead with the projects.

Uganda

Land Tenure and Natural Resource Management

Since no part of the Serengeti ecosystem is in Uganda, there is no need to consider its history and governance in such detail as Tanzania and Kenya. There are, after all, no wildebeest in Uganda. Uganda's importance to the Serengeti ecosystem stems from its strategic location on Lake Victoria and the headwaters of the Nile. Those who wish more background can find an excellent and concise account in the U.S. State Department Background Note on Uganda issued by the Bureau of African Affairs (November 28, 2012).

Since gaining independence from Britain in 1962, Uganda has been ruled by a succession of Big Men (a.k.a. dictators), beginning with Apollo Milton Obote (1962–71), Idi Amin (1971–79), Obote again (1980–85), Gen. Tito Okello (1985–86), and Yoweri Musaveni (1986–present). Although Idi Amin achieved the greatest notoriety, the others also had murderous regimes that attempted ethnic cleansing of competing tribes and kingdoms (e.g., Buganda).

It was not until 1995, when a new constitution established Uganda as a republic with executive, legislative, and judicial branches, that property rights were taken back from the central government and restored to the people. The Land and Environment Act (Chapter 15 of the Constitution) and the Land Act of 1998 reinstalled the systems of land tenure that were in existence before independence. These were stated as customary tenure, mailo tenure (a system peculiar to Buganda), freehold tenure, and leasehold tenure.

After decades of internal strife, Uganda has experienced more than twenty years of relative political stability and economic growth. Since assuming power in early 1986, Museveni's government has taken important steps toward economic rehabilitation and has adopted policies that have promoted rapid economic development. An estimated 2.5 billion barrels of recoverable oil near the Congo border in the Albertine Rift will make Uganda one of sub-Saharan Africa's top oil producers, potentially doubling current government revenues within a decade. However, it faces challenges, including continued rapid population growth, human rights problems, inflation, corruption, and flawed 2011 presidential and parliamentary elections (U.S. State Dept. Background Note: Uganda).

Development of hydroelectric power has been and still is one of Uganda's top priorities. Construction of the Owens Falls dam in 1959 transformed Lake Victoria into a giant reservoir. To reassure Egypt that operation of the dam would not reduce the flow of water from the Nile headwaters to the lower Nile River, Uganda signed several agreements to maintain the outflow from the dam to an "agreed curve" determined by natural fluctuations in the lake's level. But in 2000, Uganda completed a second dam, Kiria, near Jinja, and began operating both dams in parallel. That made it impossible to continue honoring the agreement; Uganda has been in violation since 2002. As a result, discharge down the White Nile River increased by 30 to 50 percent, and the level of Lake Victoria dropped by 2 m to its lowest level in fifty years.

The lowered water level has destroyed vast areas of papyrus swamps, important because they take up excess nutrients and provide vital habitat for juvenile tilapia. As a result, fish numbers are decreasing, and the price of fish is increasing.[53] (When fish become scarce or expensive, bushmeat poaching in the park increases.)[27] The possible collapse of the fishing industry would affect the millions of people who reside in the Lake Victoria Basin, estimated at about one-third of the total population of Kenya, Uganda, and Tanzania.

Meanwhile, a third dam on the Victoria Nile is under construction at the Bujagali Falls.

ADDITIONAL CHALLENGES

Climate Change

Those who doubt that global warming is affecting the East African climate need to find an alternative explanation for the disappearing glaciers on Mounts Kilimanjaro and Kenya and Uganda's Rwenzori Range. Pollution of the atmosphere with CO_2 and other greenhouse gases and/or by changes in solar radiation is expected to increase average global temperature by 2–5°C by midcentury.[54] Consequently, the Serengeti is almost certain to be subjected to greatly increased atmospheric CO_2. Most global climate models (GCMs) predict that East African rainfall will increase by 2–5 percent (25–50 mm) by 2050, in the form of longer, wetter rainy seasons. The underlying factor in the prediction is an expected 2°C warming of the Indian Ocean. Warmer sea surface would extend the wet season by increasing the moisture content of the trade winds reaching the interior and intensifying the intertropical convergence zone.[54]

However, averaged rainfall records of weather stations across Serengeti NP dating from the 1960s do not show the expected increase, despite warming of the Indian Ocean. Annual and wet-season rainfall did increase in the Serengeti in the first half of the twentieth century, but since 1960 annual and rainy season precipitation has declined significantly, this despite a slow increase in dry-season rainfall throughout the century.

The discrepancy between the changes predicted by climate models and the reality check of actual records demonstrates how the interaction of the various forces that affect climate produce widely varying results, as discussed in chapter 2 on the African tropics.

"A key element of climate change," writes Ritchie,[54] "is uncertainty—both the uncertainty of the net effects of different global change factors, and the risks imposed if global changes lead to increased variability in climate" (183). It is also possible, he points out, that a warmer ocean will weaken onshore winds by reducing the temperature gradient to the extent that moist air will not reach as far inland as the Serengeti. So less rain will fall in the rainy season and the climate will become more arid.

Climatic variability within the same region can be illustrated by the differences in recorded temperature between Amboseli NP and Serengeti NP. Between 1981 and 1990 (the only years with available Serengeti NP temperature records), Serengeti temperature decreased while Amboseli temperature steadily increased. The lower elevation and reduced woodland of Amboseli can at least partially explain the difference, while different climate models predict little temperature change in the East African highlands.

The other global scenario has the Serengeti responding to major climate changes by becoming drier, with decreased average wet-season rainfall and greater variability, with longer stretches of consecutive wet and dry years. Elevated CO_2 is predicted in both scenarios. These changes are likely to have a major impact on the ecosystem, on biodiversity, on the migration, on human welfare, and other consequences highlighted here and in various publications of Serengeti researchers. Taken together, the expected change in CO_2, rainfall, and temperature will probably reduce primary productivity by 20 percent.[55]

Intact Savanna Ecosystems as Major Carbon Sinks

But on the plus side, measurements of soil organic carbon in Serengeti NP indicate that grasslands have the potential to sequester globally significant quantities of greenhouse gases. They may accordingly acquire monetary value as international efforts to reduce atmospheric pollutants intensify.

Utilizing exclosure plots established by McNaughton (see chapters 6 and 9), scientists measured the effects of grazing on soil organic carbon (SOC) in a natural tropical savanna. (Grazing was already known to have strong effects but whether positive or negative depended on grazing intensity and factors such as climate, soils, fire, and prevailing land use.) They found that intermediate grazing intensity, approximating the overall average for the Serengeti, favors deposit of carbon in the soil, whereas undergrazing results in carbon loss when the remaining tall grass fuels dry-season fires.[55]

The following statement by Ritchie[55] highlights the critical contribution of wildebeest and other grazers in maintaining savanna ecosystems: "A simple model of plant production linked to losses of carbon from fire, plant inputs of stable carbon (lignin and cellulose), and inputs of carbon through dung explains more than 80% of the variation in observed soil carbon stocks in the Serengeti, and supports the role of grazers as a primary driver of soil carbon."

Invasive Species

Tanzania, like just about every country on the globe, has been invaded by alien plant species. Human immigrants have been major vectors, bringing exotics from their home countries to plant around their new homes as ornamentals, flower borders or hedges, and vegetables. Most of the other alien invaders occur along rivers, roadways, and in settlements. Many alien species introduced outside their natural range are harmless. But others can change the functioning of ecosystems, for example, by altering the fire regime, nutrient cycling, and hydrology, and can cause substantial harm to biodiversity and human livelihoods. Invasive alien species are rightly regarded as the second greatest threat to biological diversity globally, after habitat fragmentation. "The harsh reality concerning the problem of invasive alien species is that it becomes far more expensive to control invasions the longer they are left, and the chances of controlling them effectively diminishes simultaneously over time."[56]

Protected areas, including World Heritage Sites like the Serengeti and Ngorongoro have their share of invasives. A researcher from the Tanzania Wildlife Research Institute found eleven invasive species along roadways in Serengeti NP:

Amaranthus hybridus, slender pigweed

Euphorbia tirucalli, pencil tree hedges (native?)

Opuntia vulgaris, prickly pear, Mexico

Argemone mexicana, Mexican prickly poppy, poisonous to grazing animals

Datura stramonium, jimsonweed, annual herb, Asia, poisonous

Tagetes minuta, marigold (refreshing drink in South America)

Cassia occidentalis, coffee weed, stinking weed, medicinal

Xanthium strumarium, India, annual herb (medicinal), reported as fatal to cattle and pigs

Bidens pilosa (tropical and Central America), a hardy weed capable of invading a vast range of habitats, known to invade grassland, etc., and to significantly reduce crop yields

Ricinus communis, Euphorbiaceae, castor oil plant, Asia and Africa, highly toxic

Lantana camara, Verbenaceae, southeastern U.S. origin, highly toxic

Senna didymobotrya, candelabra tree, perennial shrub, pea family

Gomphocarpus kaesneri, Asclepiadaceae (Daily News Co., Tanzania, December 11, 2010)

Pistia stratiotes, Nile cabbage, South America, major weed of lakes, dams, ponds, and slow-moving waterways in the Serengeti region

Two others have been recorded in the NCA:[57]

Acacia mearnsii, black wattle

Mimosa pigra, giant sensitive plant

These two, along with *Lantana camara,* are included in the global list of the hundred worst invasive alien species (www.issg.org/database).[58]

But by far the biggest threat to Serengeti rangelands is the highly invasive alien weed, *Parthenium hysterophorus,* a South American annual herb of the aster family (Asteraceae) that is considered one of the world's seven most devastating and hazardous weeds (fig. 12.3). It was seen growing in the Masai Mara Reserve.

Parthenium, which can grow from seed to maturity in 4–6 weeks and produce 10,000–25,000 seeds, is known to be allelopathic (i.e., it produces chemicals that inhibit the growth of other plants). If it invades natural pasture it can reduce the amount of available forage to such an

FIGURE 12.3. *Parthenium hysterophorus*, a toxic South American weed that can destroy grassland and crops, is gaining a foothold in the Masai Mara Reserve.

extent that carrying capacities of grazing animals can be reduced by up to 90 percent. If allowed to grow without weeding it can reduce yields of crops, such as sorghum, by up to 97 percent. It is also toxic, which means that animals will not eat it unless they are starving or stressed, with fatal consequences. This weed also has impacts on human health: many people who come into direct contact with the plant can develop severe skin allergies (dermatitis), and pollen production by the plant can result in respiratory problems.[56]

The implications for wildlife conservation in the Serengeti ecosystem are potentially extremely serious. Allan Kijazi, director general of TAN-APA, said that TANAPA and their Kenya counterpart are working on ways to eliminate the weeds from the ecosystem. "Unless action is taken immediately to eradicate known infestations in the Serengeti–Masai Mara ecosystem," Kijazi said, "it is not unrealistic to expect a drastic reduction in wildlife populations as the *Parthenium* population rapidly expands" (*East African*, November 24, 2010).

It is therefore possible for a little green plant to transform one of the greatest wildlife spectacles on earth. Fortunately, the invasion is still in its early stages. If sustained and effective eradication measures are undertaken in the near future, it can be stopped.[56]

IS THE SERENGETI ECOSYSTEM UP TO ALL THESE CHALLENGES?

Hilborn, Sinclair, and Fryxell,[17] who between them have spent a century studying Serengeti ecology, have faith in the resilience of the ecosystem. They point out that it has absorbed and survived a wide range of perturbing events, notably long-term changes from woodland to grassland and back again, with consequent responses of the fauna. Following the rinderpest pandemic of the 1890s, woodland replaced savanna and plains and elephants reappeared. In the 1960s and 1970s, coinciding with higher dry-season rainfall, woodlands opened up and the wildebeest population doubled and redoubled. Then the herbivores left so little long grass that fires became infrequent and poachers decimated the elephant population. This caused the balance to shift back to woodland in the areas of higher rainfall.

The sheer size of the ecosystem and the extraordinary diversity of habitats and wildlife are key to its resilience. This resilience would be greatly compromised if the two million wildebeest and zebra could not migrate from the Serengeti plains to the Masai Mara in their quest for the most nutritious pastures and access to the perennial water source of the Mara River. In addition to the effect on the migration, the change would have cascading effects on the herbivore guild that the migration spares from unremitting competition (as reviewed in chapter 9).

In their conclusions, the authors[17] ascribe the resilience of the Serengeti in part to the hands-off policy of park managers, which has let the system move through a range of different states with minimal interference. "Do not try to hold the system in one state," the authors advise. "Serengeti is largely an unmanaged ecosystem—the only major 'management' is protection from poaching, grazing, and encroachment" (437).

The Need for Regional Cooperation

To state the obvious, the future of the Serengeti ecosystem depends on the willingness of Tanzania, Kenya, and Uganda to put the survival of a unique landscape that is not only their own heritage but also that of the world ahead of national development policies that will destroy it. The framework for their cooperation is the East African Community (EAC).

Originally founded in 1967, the EAC became defunct in 1977 but was officially revived in 2000 by agreement of the presidents of Uganda, Kenya, and Tanzania. Membership was later extended to include Burundi and Rwanda. In 2012, the East African Legislative Assembly, the legislative arm of the EAC, established the Trans-boundary Ecosystems Management Commission, which would ensure that one country could not make unilateral decisions that would affect the ecosystem of another without a rigorous study and process. It emphasized the need for an environmental impact assessment (EIA) of projects with transboundary impacts and a key role by the commission in the approval process.

The bill needed the approval of all EAC presidents, but Tanzania's president refused to sign. This added to suspicions that Tanzania in fact has not given up plans to construct a commercial thoroughfare across the park, especially considering that upgrading of roads is proceeding on both sides. It still has not taken up the World Bank's offer to help finance the southern highway outside the park, or accepted the German government's offer to participate in the development of the southern route. It even remains unclear whether the southern route concept has been definitely accepted as an alternative to the Serengeti highway. No report has appeared in the Tanzania press of Germany's offer of 23.5 million euros to build local roads and fund projects to relieve poverty in neighboring communities in exchange for rerouting the Serengeti highway.

Subsequently, Tanzania was taken to the East African Court of Justice by the Africa Network of Animal Welfare (ANAW), a local Kenya NGO, challenging the government's decision to build a highway across Serengeti NP on the grounds that it would have far-reaching consequences for the Serengeti-Mara ecosystem.

On March 21, 2012, the *Daily Nation* reported that the case would move to trial. A judge with the EAC threw out the Tanzania attorney general's argument that the issue was a national one and not regional. The judge said in his ruling, quoted in the newspaper, "The court has jurisdiction to hear environmental disputes which directly affect the ecosystem and touch on the sustainable utilization of the natural resources."

If ANAW wins this case, the prospects for preserving the integrity of the Serengeti ecosystem and other ecosystems shared by EAC members will be greatly enhanced. Tanzania, in turn, could challenge Kenya's right to diminish and pollute the Mara River. Either or both countries could take Uganda to court to prevent degradation of the Lake Victoria environment by lowering the water level. This could be done through the Lake Victoria Basin Commission (LVBC), a specialized institution of the EAC responsible for coordinating the sustainable development agenda of the Lake Victoria Basin.

Other international policies and agreements potentially applicable to the greater Serengeti ecosystem,[21] include the following:

African Convention on the Conservation of Nature and Natural Resources 1968: preservation of wildlife and other natural resources

Ramsar Convention 1971: wetland and migratory waterfowl conservation

Convention on International Trade in Endangered Species (CITES) 1973: governs international trade of endangered species

Bonn Convention on Migratory Species 1979

World Heritage Listing 1979 and 1981 (UNESCO): focuses on preserving world/cultural heritages.

Man and Biosphere Reserve 1981: goal to perpetuate earth's living organisms through collaboration between the world's governments and NGOs

Lusaka Agreement 1994: reduce and eliminate illegal trade in wild flora and fauna

African Eurasian Migratory Water Bird Agreement (AEWA)

Monitoring the Illegal Killing of Elephants (MIKE)

It is doubtful that any of these organizations can exert as much influence on the conservation of the Serengeti as the EAC. This is not to say that international organizations should not follow events affecting the Serengeti ecosystem and exert as much influence as possible. Let us not forget that the rallying of international opposition to the proposed Serengeti road by hundreds of scientists, associations of tour operators, and developments publicized by Serengeti Watch and other social media eventually caused Tanzania to say it would give up the highway and accept the alternative route.

Suppose the Serengeti Ecosystem Became Part of an Enlarged Transfrontier Conservation Area

Missing from the above list are transfrontier parks or transfrontier conservation areas, mentioned in the accounts of migratory wildebeest populations in southern Africa (chapter 4). Incorporating huge landscapes including both protected and unprotected lands, they offer the best hope for restoring the fragmented ecosystems of migratory populations. The only East African example was the agreement by Mozambique and Tanzania to establish and maintain a corridor reconnecting the Selous Game Reserve to the Niassa Game Reserve—unfortunately discontinued in 2012 when Mozambique failed to renew the contract of the Niassa Reserve management. Also known as Peace Parks, transfrontier parks are intended not only for the conservation and sustainable use of biological and cultural resources; they also have the objective of facilitating and promoting regional peace and cooperation and socioeconomic development. (See Peace Parks Foundation website.)

The Serengeti ecosystem should be transformed into a transfrontier park/conservation area. Ideally, it would incorporate most of the savanna habitat lost from the wider ecosystem, including the pastoral Masai along with their present and previous rangelands, encompassing group ranches, the Loita Plains, maybe even the Isiria Escarpment. Establishing and managing such a park/conservation area will be a challenging task, as it would include areas with different systems of land tenure, types of land use, and degrees of development. Furthermore, the area concerned is governed and managed by two independent nations, an array of government ministries and departments, and a number of cooperative and private enterprises.

Getting the three nations involved to agree to an international park/conservation area stands as a major hurdle. But making a transfrontier park incorporating the Serengeti ecosystem could be so much simpler than connecting countries with different languages and governance. Another problem is that much of the original Serengeti ecosystem within Kenya has been transformed into wheat fields; formerly communal lands have been privatized. Landowners will be reluctant to allow wildlife back onto these properties as long as agriculture and commercial development provide far greater returns[38].

There is hope, however. For instance, wildlife could become more valuable than other uses if landowners received a fair share of the income from ecotourism in the transfrontier park. They also may find their land more suited to pastoralism than agriculture if the regional climate becomes more arid (as shown by some projections).

The Masai are well along in the process of giving up a pastoral existence in favor of living in permanent settlements with educational and medical facilities. Those persisting in a pastoral existence generally have fewer cattle and live below the poverty line. But this could change should ranching cattle, sheep, and goats to satisfy the huge potential market for meat (now met in large part by poached bushmeat) become more profitable. With this in mind, pastoralists might be reintroduced into the Serengeti ecosystem by being granted access to at least some of their former rangelands. They would, of course, have to accept limitations on the number of livestock and to share the range with wildlife, as always in the past. As the Masai take pride in their best-endowed livestock, the key to changing the paradigm from quantity to quality is already there.

The anthropologists and sociologists who have studied pastoralism in East Africa back proposals for wildlife and livestock to coexist even in protected areas.[39,59,60] As they shared the savannas for thousands of years to their mutual benefit, why couldn't they do it again in the transfrontier park? Transhumance worked and can work again without degrading the habitat in extensive and diverse savanna landscapes.

Another example of problems that need addressing is the notable lack of cooperation between Kenyan and Tanzanian wildlife authorities as evinced by the fact that—unless it has happened very recently—the respective management teams have yet to coordinate aerial counts of Masai Mara and Serengeti wildlife. Surely cooperation could be greatly improved given the necessary incentives (or better communication?).

So, yes, it will be a challenging task. Nevertheless, the establishment of a transfrontier park is important—perhaps even necessary—to the survival of the Serengeti ecosystem.

THE WAY FORWARD—OR BACKWARD?

Creating a transborder conservation area may offer the best hope of perpetuating the Serengeti ecosystem, but making the necessary arrangements for joint management are unlikely in the immediate future. Unless the EAC members accept the jurisdiction and decisions of the East African Court of Justice, what are the chances the Kenya government will restore the Mara River's flow by limiting irrigation offtake and reforesting the Mara catchment forest? Can Uganda be prevailed upon to restore sustainable flows from Lake Victoria at the cost of reducing the hydroelectric power produced by the three dams? Does irrevocably damaging the lake's natural resources outweigh producing the electricity required for development?

Consider the consequences if all these initiatives came to naught. What a loss to world knowledge if research pursued for over six decades in the last great savanna ecosystem stops just as Serengeti researchers are beginning to comprehend its extraordinary biological biodiversity. The realization that a landscape epitomizing the savannas in which our species evolved can be destroyed in this century without proactive intervention of *Homo sapiens* is profoundly depressing.

Also consider what has happened to other major migratory ecosystems at the hand of humankind. In Africa, the great migratory populations of the South African highveld, the Kalahari, the lechwe of the Kafue Flats and Chobe NP, the wildebeest hordes that thronged the plains between Mount Meru and the Gregory Rift Valley and the Athi-Kapiti Plains, are no more. In the Northern Savanna, civil war has spared the white-eared kob and tiang migration in Southern Sudan, but the populations of addax, scimitar-horned oryx, and dama gazelle have been virtually exterminated in the Sahel and Sahara biomes. In North America, an estimated 30 million bison that populated the prairies and Great Plains were reduced to hundreds before 1900. In Eurasia, only one of the ungulates that existed in millions, the Mongolian gazelle, still migrates in sizable numbers on the Asian Steppe, though threatened by developments; the once huge saiga population has been reduced by 90 percent since the breakup of the Soviet Union.

And what has supplanted them? Domesticated cattle, sheep, and goats so unfit to compete with native African and North American ruminants that they cannot survive without human intervention. The squandering of these once-abundant, diverse wildlife populations stands out as one of the most self-destructive acts of the human race.

Viewed from this perspective, survival of the Serengeti ecosystem seems miraculous and even more deserving of protection from the dangers that threaten to destroy it. The whole world has a stake in keeping this greatest African savanna ecosystem alive. SERENGETI MUST NOT DIE!

REFERENCES

1. Sinclair and Arcese 1995.
2. Arcese, Hando, and Campbell 1995.
3. Sinclair 2012.
4. USAID Country Profile 2011.
5. Nshala 1999.
6. Leader-Wlliams et al. 2008.
7. Shauri and Hitchcock 1999.
8. Baldus and Cauldwell 2004.

9. Baldus, Kaggi, and Ngoti 2004.
10. Tanzania 2012.
11. Emerton and Mfunda 1999.
12. Songorwa 1999.
13. Ministry of Natural Resources and Tourism 1985.
14. Ahni Schertow 2011.
15. Majamba 2001.
16. Nelson 2005.
17. Hilborn, Sinclair, and Fryxell 2008.
18. Songorwa 2004.
19. A. Estes et al. 2012.
20. Schmitt 2010.
21. Polasky et al. 2008.
22. Tanzania National Bureau of Statistics 2002.
23. Scholte and de Groot 2010.
24. Barnett 2000.
25. Campbell, Nelson, and Loibooki 2001.
26. Campbell and Borner 1995.
27. Rentsch 2012.
28. Pascual and Hilborn 1995.
29. Mduma, Sinclair, and Hilborn 1999.
30. Estes, Raghunathan, and Van Vleck 2008.
31. Fitzgibbon 1998.
32. UNESCO 2006.
33. Campbell and Hofer 1995.
34. Hofer et al. 1996.
35. Turner and Jackman 1987.
36. Thirgood et al. 2008.
37. Dobson et al. 2010.
38. Norton-Griffiths et al. 2005.
39. Homewood et al. 2001.
40. Ottichilo, de Leeuw, and Prins 2001.
41. Norton-Griffiths 1995.
42. Serneels and Lambin 2001b.
43. Ogutu et al. 2009.
44. Wanjala 2000.
45. Campbell 1984.
46. Grandin 1986.
47. Rutten 1992.
48. Kimani and Picard 1998.
49. Galaty. 1994.
50. Gereta et al. 2002.
51. Wolanski et al. 1999.
52. Wolanski and Gereta 2001.
53. Mnaya et al. 2011.
54. Ritchie 2008.
55. Ritchie in press.

56. Clark and Lotter 2008.
57. PAMS Foundation 2009.
58. IUCN Global Invasive Species Database.
59. Reid 2012.
60. Marshall 1990.

Bibliography

Ahni Schertow, J. 2011. "Tanzania: Pastoralists Refuse to Leave Maswa Game Reserve." *International Cry* (May). http://intercontinentalcry.org/tanzania-pastoralists-refuse-to-leave-maswa-game-reserve/.

Alden, P., R.D. Estes, D. Schlitter, and B. McBride. 1995. *The National Audubon Society Field Guide to African Wildlife.* New York: Knopf Doubleday Publishing Group.

Alexander, R.McN. 1977. "Allometry of the Limbs of Antelopes (Bovidae)." *Journal of Zoology* 183, no. 1: 125–46.

Amiyo, A.T. 2006. "Ngorongoro Crater Rangelands: Condition, Management and Monitoring." M.Sc. thesis, University of Kwazulu-Natal.

Anderson, G.D. 1963. "Some Weakly Developed Soils of the Eastern Serengeti Plains, Tanganyika." *African Soils* 8: 339–47.

———. 1965. "Estimates of the Potential Productivity of Some Pastures in the Mbulu Highlands of Tanganyika by Application of Fertilizers." *East African Agricultural and Forestry Journal* 30, no. 3: 206–18.

Anderson, G.D., and L.M. Talbot. 1965. "Soil Factors Affecting the Distribution of the Grassland Types and Their Utilization by Wild Animals on the Serengeti Plains, Tanganyika." *Journal of Ecology* 53, no. 1: 33–56.

Anderson, T.M., J. Dempewolf, K.L. Metzger, D.N. Reed, and S. Serneels. 2008. "Generation and Maintenance of Heterogeneity in the Serengeti Ecosystem." In *Serengeti III: Human Impacts on Ecosystem Dynamics,* edited by A.R.E. Sinclair, C. Packer, S.A.R. Mduma, and J.M. Fryxell, 135–82. Chicago: University of Chicago Press.

Anderson, T.M., J.G.C. Hopcraft, S. Eby, M. Ritchie, J.B. Grace, and H. Olff. 2010. "Landscape-Scale Analyses Suggest Both Nutrient and Anti-predator Advantages to Serengeti Herbivore Hotspots." *Ecology* 91, no. 5: 1519–29.

Arcese, P., J. Hando, and K. Campbell. 1995. "Historical and Present-Day Anti-Poaching Efforts in Serengeti." In *Serengeti II: Dynamics, Management, and Conservation of an Ecosystem,* edited by A. R. E. Sinclair and P. Arcese, 506–33. Chicago: University of Chicago Press.

Arctander, P., C. Johansen, and M. A. Coutellec-Vreto. 1999. "Phylogeography of Three Closely Related African Bovids (Tribe Alcelaphini)." *Molecular Biology and Evolution* 16, no. 12: 1724–39.

Arsenault, R., and N. Owen-Smith. 2002. "Facilitation versus Competition in Grazing Herbivore Assemblages." *Oikos* 97, no. 3: 313–18.

Attwell, C. A. M. 1977. "Reproduction and Population Ecology of the Blue Wildebeest *Connochaetes t. taurinus* in Zululand." PhD dissertation, University of Natal.

———. 1982. "Population Ecology of the Blue Wildebeest *Connochaetes taurinus taurinus* in Zululand, South Africa." *African Journal of Ecology* 20, no. 3: 147–68.

Attwell, C. A. M., and J. Hanks. n.d. "Reproduction of the Blue Wildebeest *Connochaetes taurinus taurinus* in Zululand, South Africa." *Säugetierkundliche Mitteilungen* 28, no. 4: 264–81.

Augustine, D. J., and S. J. McNaughton. 2004. "Regulation of Shrub Dynamics by Native Browsing Ungulates on East African Rangeland." *Journal of Applied Ecology* 41, no. 1: 45–58.

Baldus, R. D. 2009. *A Practical Summary of Experiences after Three Decades of Community-Based Wildlife Conservation in Africa: "What Are the Lessons Learnt?"* CIC Technical Series Publication 5. Rome: CIC and FAO.

Baldus, R. D., and A. E. Cauldwell. 2004. "Tourist Hunting and Its Role in Development of Wildlife Management Areas in Tanzania." *Game and Wildlife Science* 21, no. 4: 591–614.

Baldus, R. D., G. T. Kaggi, and P. M. Ngoti. 2004. "Community Based Conservation (CBC): Where Are We Now? Where Are We Going?" *Kakakunoa* 35: 20–22.

Balmford, A., J. C. Deutsch, R. J. C. Nefdt, and T. Clutton-Brock. 1993. "Testing Hotspot Models of Lek Evolution: Data from Three Species of Ungulates." *Behavioral Ecology and Sociobiology* 33, no. 1: 57–65.

Barnett, R. 2000. *Food for Thought: The Utilization of Wild Meat in Eastern and Southern Africa.* Nairobi: Traffic East/Southern Africa.

Bell, R. H. V. 1970. "The Use of the Herb Layer by Grazing Ungulates in the Serengeti." In *Animal Populations in Relation to Their Food Resources: A Symposium of the British Ecological Society, Aberdeen, 24–28 March 1969,* edited by A. Watson, 111–24. Oxford: Blackwell Scientific.

———. 1971. "A Grazing Ecosystem in the Serengeti." *Scientific American* 225, no. 1: 86–93.

Belsky, A. J. 1984. "Small-Scale Pattern in Grassland Communities in the Serengeti National Park, Tanzania." *Vegetatio* 55, no. 3: 141–51.

Berry, H. H. 1997. "Aspects of Wildebeest *Connochaetes taurinus* Ecology in the Etosha National Park—A Synthesis for Future Management." *Madoqua* 20, no. 1: 137–48.

———. 1981. "Population Structure, Mortality Patterns, and a Predictive Model for Estimating Future Trends in Wildebeest Numbers in the Etosha National Park South Africa." *Madoqua* 12, no. 4: 255–66.

Bolger, D. T., W. D. Newmark, T. A. Morrison, and D. F. Doak. 2008. "The Need for Integrative Approaches to Understand and Conserve Migratory Ungulates." *Ecology Letters* 11, no. 1: 63–77.

Bond, W. J. 2008. "What Limits Trees in C4 Grasslands and Savannas?" *Evolution and Systematics* 39: 641–59.

Bond, W. J., F. I. Woodward, and G. F. Midgley. 2005. "The Global Distribution of Ecosystems in a World Without Fire." *New Phytology* 165, no. 2: 525–37.

Bourlière, F. 1961. "Symposium sur les deplacements saisonniers des animaux." *Revue Suisse de Zoologie* 68, no. 2: 139–43.

Bourlière, F., and J. Verschuren. 1960. *Introduction à l'écologie des ongulés du Parc National Albert.* Vol. fasc. 1. Exploration du Parc National Albert, Mission F. Bourlière et J. Verschuren. Brussels: Institut des parcs nationaux du Congo Belge.

Breman, H., and C. T. de Wit. 1983. "Rangeland Productivity and Exploitation in the Sahel." *Science* 221, no. 4618: 1341–47.

Briske, D. D., S. D. Fuhlendorf, and F. E. Smeins. 2006. "A Unified Framework for Assessment and Application of Ecological Thresholds." *Rangeland Ecology & Management* 59, no. 3: 225–36.

Bro-Jorgensen, J. 2002. "Overt Female Mate Competition and Preference for Central Males in a Lekking Antelope." *Proceedings of the National Academy of Sciences* 99, no. 14: 9290–93.

Brooks, A. C. 1961. *A Study of the Thomson's Gazelle* (Gazella thomsonii Gunther) *in Tanganyika.* Colonial Research Publication 25. London: HMSO.

Brown, L. 1965. *Africa: A Natural History.* London: Hamish Hamilton.

Butynski, T. 2013. "*Beatragus hunteri* (Sclater)." In *Mammals of Africa,* vol. 6, edited by J. Kingdon and M. Hoffman, 491–95. London: Bloomsbury Press.

Campbell, D. J. 1984. "Response to Drought among Farmers and Herders in Southern Kajiado District." *Human Ecology* 12: 35–61.

Campbell, K., and M. Borner. 1995. "Population Trends and Distribution of Serengeti Herbivores: Implications for Management." In *Serengeti II: Dynamics, Management, and Conservation of an Ecosystem,* edited by A. R. E. Sinclair and P. Arcese, 117–45. Chicago: University of Chicago Press.

Campbell, K., and H. Hofer. 1995. "People and Wildlife: Spatial Dynamics and Zones of Interaction." In *Serengeti II: Dynamics, Management, and Conservation of an Ecosystem,* edited by A. R. E. Sinclair and P. Arcese, 534–70. Chicago: University of Chicago Press.

Campbell, K. L. I., V. Nelson, and M. Loibooki. 2001. *Sustainable Use of Wildland Resources, Ecological, Economic, and Social Interactions: An Analysis of Illegal Hunting of Wildlife in Serengeti National Park, Tanzania.* Chatham, UK: Department for International Development (DFID) Animal

Health Programme and Livestock Production Programmes, Natural Resources Institute.

Caro, T. M., and C. D. Fitzgibbon. 1992. "Large Carnivores and Their Prey: The Quick and the Dead." In *Natural Enemies: The Population Biology of Predators, Parasites and Diseases,* edited by M. J. Crawley, 117–42. Oxford: Blackwell Scientific.

Chapman, D. I. 1972. "Seasonal Changes in the Gonads and Accessory Glands of Male Mammals." *Mammal Review* 1, no. 7–8: 231–48.

Chardonnet, P., and B. Chardonnet. 2004. *Antelope Survey Update.* IUCN/SSC Antelope Specialist Group Report 9. Paris: *Fondation internationale pour la sauvegarde de la faune.*

Clark, K., and W. Lotter. 2008. "Stop the Invasion before It Is Too Late." Kakakuona no. 48. PAMS Foundation.

Clutton-Brock, T. H. 1994. "The Costs of Sex." In *The Differences between the Sexes,* edited by R. V. Short and E. Balaban, 347–62. Cambridge: Cambridge University Press.

Clutton-Brock, T. H., J. C. Deutsch, and R. J. C. Nefdt. 1993. "The Evolution of Ungulate Leks." *Animal Behaviour* 46, no. 6: 1121–38.

Conway, G. 2009. "The Science of Climate Change in Africa: Impacts and Adaptation." Discussion Paper No. 1, presented at the Grantham Institute for Climate Change.

Cowie, I. 2004. "Update on Nairobi National Park." *Gnusletter* 23, no. 1: 16.

———. 2005. "Nairobi National Park." *Gnusletter* 23, no. 2: 17–19.

Cumming, D. H. M. 1999. *Study on the Development of Transboundary Natural Resource Management Areas in Southern Africa. Environmental Context: Natural Resources, Land Use, and Conservation.* Washington DC: Biodiversity Support Program.

Darling, F. F. 1960a. *An Ecological Reconnaissance of the Mara Plains in Kenya Colony.* Wildlife Monographs 5. Washington DC: Wildlife Society.

———. 1960b. *Wild Life in an African Territory: A Study Made for the Game and Tsetse Control Dept. of Northern Rhodesia.* London: Oxford University Press.

Darwin, C. 1871. *The Descent of Man and Selection in Relation to Sex.* London: Murray.

Dawson, J. B. 1992. "Neogene Tectonics and Volcanicity in the North Tanzanian Sector of the Gregory Rift Valley: Contrasts with the Kenyan Sector." *Tectonophysics* 204: 81–92.

De Boer, W. F., and H. H. T. Prins. 1990. "Large Herbivores That Strive Mightily but Eat and Drink as Friends." *Oecologia* 82: 264–74.

Demment, M. W., and P. J. Van Soest. 1985. "A Nutritional Explanation for Body-Size Patterns of Ruminant and Nonruminant Herbivores." *American Naturalist* 125, no. 5: 641–72.

Dempewolf, J., S. Trigg, R. S. DeFries, and S. Eby. 2007. "Burned-Area Mapping of the Serengeti-Mara Region Using MODIS Reflectance Data." *Geoscience and Remote Sensing Letters, IEEE* 4, no. 2: 312–16.

Dobson, A. P., M. Borner, A. R. E. Sinclair, P. J. Hudson, T. M. Anderson, G. Bigurube, T. B. B. Davenport, et al. 2010. "Road Will Ruin Serengeti." *Nature* 467, no. 7313: 272–73. doi:10.1038/467272a.

Dobzhansky, T. 1937. "Isolating Mechanisms." In *Genetics and the Origin of Species,* 228–58. New York: Columbia University Press.

Dougall, H.W. 1963. "Changes in the Chemical Composition of Plants during Growth in the Field." *East African Agricultural and Forestry Journal* 28: 182–89.

Dougall, H.W., V.M. Drysdale, and P.E. Glover. 1964. "The Chemical Composition of Kenya Browse and Pasture Herbage." *East African Wildlife Journal* 2: 86–121.

Dublin, H.T. 1986. "Decline of the Mara Woodlands: The Role of Fire and Elephants." PhD dissertation, University of British Columbia.

———. 1995. "Vegetation Dynamics in the Serengeti-Mara Ecosystem: The Role of Elephants, Fire and Other Factors." In *Serengeti II: Dynamics, Management, and Conservation of an Ecosystem,* edited by A.R.E. Sinclair and P. Arcese, 71–90. Chicago: University of Chicago Press.

Dublin, H.T., A.R.E. Sinclair, S. Boutin, E. Anderson, M. Jago, and P. Arcese. 1990. "Does Competition Regulate Ungulate Populations—Further Evidence from Serengeti, Tanzania." *Oecologia* 82, no. 2: 283–88.

Dublin, H.T., A.R.E. Sinclair, and J. McGlade. 1990. "Elephants and Fire as Causes of Multiple Stable States in the Serengeti-Mara Woodlands." *Journal of Animal Ecology* 59, no. 3: 1147–64.

Duncan, P. 1975. "Topi and Their Food Supply." DPhil thesis, University of Nairobi.

DuPlessis, S.S. 1972. *Ecology of Blesbok with Special Reference to Productivity.* Wildlife Monographs 30. Washington DC: Wildlife Society.

East, R. 1999. *African Antelope Database 1998.* Occasional Paper of the IUCN Species Survival Commission 21. Cambridge: International Union for Conservation of Nature.

Ehret, C. 2002. *The Civilizations of Africa: A History to 1800.* Charlottesville: University Press of Virginia.

Emerton, L., and I. Mfunda. 1999. *Making Wildlife Economically Viable for Communities Living around the Western Serengeti, Tanzania.* Evaluating Eden Discussion Paper 1. London: International Institute for Environment and Development, Biodiversity and Livelihoods Group.

Epp, H. 1980. *A Natural Resources Survey of Serengeti National Park Tanzania.* Serengeti Research Institute Publications 237. Arusha: Serengeti Research Institute.

Estes, A.B., T. Kuemmerle, H. Kushnir, V.C. Radeloff, and H.H. Shugart. 2012. "Land-Cover Change and Human Population Trends in the Serengeti Ecosystem from 1984–2003." *Biological Conservation* 147, no. 1: 255–63.

Estes, R.D. 1963. *Second Quarterly Report to National Geographic Committee for Research and Exploration.* Washington DC: National Geographic Society.

———. 1966. "Behavior and Life History of the Wildebeest (*Connochaetes taurinus* Burchell)." *Nature* 212: 999–1000.

———. 1967a. "The Comparative Behavior of Grant's and Thomson's Gazelles." *Journal of Mammalogy* 48, no. 2: 189–209.

———. 1967b. "Predators and Scavengers, Pt. 2." *Natural History* 76, no. 3: 38–47.

———. 1969. "Territorial Behavior of the Wildebeest (*Connochaetes taurinus* Burchell, 1823)." *Zeitschrift für Tierpsychologie* 26: 284–370.

———. 1974. "Social Organization of the African Bovidae." In *The Behaviour of Ungulates and Its Relation to Management: The Papers of an International Symposium Held at the University of Calgary, Alberta, Canada, 2–5 November 1971*, 166–205. IUCN Publications, n.s. 24. Morges, Switzerland: International Union for Conservation of Nature and Natural Resources.

———. 1976. "The Significance of Breeding Synchrony in the Wildebeest." *East African Wildlife Journal* 14, no. 2: 135–52.

———. 1991a. *The Behavior Guide to African Mammals: Including Hoofed Mammals, Carnivores, Primates.* Berkeley: University of California Press.

———. 1991b. "The Significance of Horns and Other Male Secondary Sexual Characters in Female Bovids." *Applied Animal Behaviour Science* 29, no. 1–4: 403–51.

———. 1993. *The Safari Companion: A Guide to Watching African Mammals: Including Hoofed Mammals, Carnivores, and Primates.* Post Mills, VT: Chelsea Green.

———. 1999a. "Hirola: Generic Status Supported by Behavioral and Physiological Evidence." *Gnusletter* 18, no. 2: 10–11.

———. 1999b. *The Safari Companion: A Guide to Watching African Mammals Including Hoofed Mammals, Carnivores, and Primates.* Rev. and expanded ed. White River Junction, VT: Chelsea Green.

———. 2000. "Evolution of Conspicuous Coloration in the Bovidae: Female Mimicry of Male Secondary Characters as Catalyst." In *Antelopes, Deer, and Relatives: Fossil Record, Behavioral Ecology, Systematics, and Conservation,* edited by E. S. Vrba and G. B. Schaller, 234–46. New Haven, CT: Yale University Press.

———. 2013. "*Connochaetes taurinus* (Burchell)." In *Mammals of Africa,* edited by J. Kingdon, D. Happold, T. Butynski, M. Hoffmann, M. Happold, and J. Kalina, 533–43. New York: Bloomsbury.

Estes, R. D., J. L. Atwood, and A. B. Estes. 2006. "Downward Trends in Ngorongoro Crater Ungulate Populations, 1986–2005: Conservation Concerns and the Need for Ecological Research." *Biological Conservation* 131, no. 1: 106–20.

Estes, R. D., and R. East. 2009. *Status of the Wildebeest (Connochaetes taurinus) in the Wild, 1967–2005.* Working Paper 37. Bronx, NY: Wildlife Conservation Society.

Estes, R. D., and R. K. Estes. 1974. "The Biology and Conservation of the Giant Sable, *Hippotragus niger variani* Thomas, 1916." *Proceedings of the Academy of Natural Sciences of Philadelphia* 26: 73–104.

———. 1979. "The Birth and Survival of Wildebeest Calves." *Zeitschrift für Tierpsychologie* 51, no. 1: 45–95.

Estes, R. D., and J. Goddard. 1967. "Prey Selection and Hunting Behavior of the African Wild Dog." *Journal of Wildlife Management* 31, no. 1: 52–70.

Estes, R. D., T. E. Raghunathan, and D. Van Vleck. 2008. "The Impact of Horning by Wildebeest on Woody Vegetation of the Serengeti Ecosystem." *Journal of Wildlife Management* 72, no. 7: 1572–78.

Estes, R. D., and R. Small. 1981. "The Large Herbivore Populations of Ngorongoro Crater." *African Journal of Ecology* 19, no. 1–2: 175–85.

Field, C. R., and R. M. Laws. 1970. "The Distribution of the Larger Herbivores in the Queen Elizabeth National Park." *Journal of Applied Ecology*, no. 7: 273–94.

Fitzgibbon, C. 1998. "The Management of Subsistence Harvesting: Behavioral Ecology of Hunters and Their Mammalian Prey." In *Behavioral Ecology and Conservation Biology*, edited by T. Caro, 449–73. Oxford: Oxford University Press.

Fosbrooke, H. A. 1963. *Ngorongoro's First Visitor: Being an Annotated and Illustrated Translation from Dr. O. Baumann's Durch Massailand Zur Nilquelle = Through Masailand to the Source of the Nile*. Translated by G. E. Organ. Ngorongoro Conservation Area Booklet. Kampala.

———. 1972. *Ngorongoro: The Eighth Wonder*. Survival Books. London: Deutsch.

Fryxell, J. M. 1991. "Forage Quality and Aggregation by Large Herbivores." *American Naturalist* 138, no. 2: 478–98.

Fryxell, J. M., P. A. Abrams, R. D. Holt, J. F. Wilmshurst, A. R. E. Sinclair, and R. Hilborn. 2008. "Spatial Dynamics and Coexistence of the Serengeti Grazer Community." In *Serengeti III: Human Impacts on Ecosystem Dynamics*, edited by A. R. E. Sinclair, C. Packer, S. A. R. Mduma, and J. M. Fryxell, 277–300. Chicago: University of Chicago Press.

Fryxell, J. M., J. F. Wilmshurst, A. R. E. Sinclair, D. T. Haydon, R. D. Holt, and P. A. Abrams. 2005. "Landscape Scale, Heterogeneity, and the Viability of Serengeti Grazers." *Ecology Letters* 8, no. 3: 328–35.

Galaty, J. G. 1994. "Rangeland Tenure and Pastoralism in Africa." In *African Pastoralist Systems: An Integrated Approach*, edited by E. Fratkin, K. A. Galvin, and E. A. Roth, 185–204. Boulder, CO: Lynne Rienner.

Gentry, A. W. 1978. "Bovidae." In *Evolution of African Mammals*, edited by V. J. Maglio and H. B. S. Cooke, 540–72. Cambridge, MA: Harvard University Press.

———. 1990. "Evolution and Dispersal of African Bovidae." In *Horns, Pronghorns, and Antlers: Evolution, Morphology, Physiology, and Social Significance*, edited by G. A. Bubenik and A. B. Bubenik, 195–227. New York: Springer-Verlag.

Georgiadis, N. 1985. "Growth Patterns, Sexual Dimorphism and Reproduction in African Ruminants." *African Journal of Ecology* 23, no. 2: 75–87.

———. 1995. "Population Structure of Wildebeest: Implications for Conservation." In *Serengeti II: Dynamics, Management, and Conservation of an Ecosystem*, edited by A. R. E. Sinclair and P. Arcese, 473–84. Chicago: University of Chicago Press.

Gereta, E., E. Wolanski, M. Borner, and S. Serneels. 2002. "Use of an Ecohydrology Model to Predict the Impact on the Serengeti Ecosystem of Deforestation, Irrigation, and the Proposed Amala Weir Water Diversion Project in Kenya." *Ecohydrology and Hydrobiology* 2, no. 1–4: 135–42.

Gerresheim, K. 1973. "Serengeti Ecosystem: Landscape Classification Units." (Map.) Tanzania LithoLtd., Arusha.

————. 1974. *The Serengeti Landscape Classification.* Serengeti Research Institute Publication 165. Nairobi: Serengeti Ecological Monitoring Programme, African Wildlife Leadership Foundation.

Gilliland, H. B. 1952. "The Vegetation of Eastern British Somaliland." *Journal of Ecology* 40, no. 1: 91–124.

Gosling, L. M. 1974. "The Social Behavior of Coke's Hartebeest (*Alcelaphus buselaphus* Cokei)." In *The Behaviour of Ungulates and Its Relation to Management: The Papers of an an International Symposium Held at the University of Calgary, Alberta, Canada, 2–5 November 1971,* edited by V. Geist and F. R. Walther, 488–511. IUCN Publication, n.s. 24. Morges, Switzerland: International Union for Conservation of Nature and Natural Resources.

————. 1982. "A Reassessment of the Function of Scent Marking in Territories." *Zeitschrift für Tierpsychologie* 60: 89–118.

————. 1986. "The Evolution of Mating Strategies in Male Antelopes." In *Ecological Aspects of Social Evolution: Birds and Mammals,* edited by D. I. Rubenstein and R. W. Wrangham, 218–44. Princeton, NJ: Princeton University Press.

————. 1991. "The Alternative Mating Strategies of Male Topi, *Damaliscus lunatus.*" *Applied Animal Behaviour Science* 29, no. 1–4: 107–20.

Grand, T. I. 1991. "Patterns of Muscular Growth in the African Bovidae." *Applied Animal Behaviour Science* 29, no. 1–4: 471–82.

Grandin, B. E. 1986. "Land Tenure, Sub-Division, and Residential Change on a Maasai Group Ranch." *Development Anthropology Network* 4, no. 2: 9–13.

Grange, S., and P. Andrews. 2006. "Bottom-Up and Top-Down Processes in African Ungulate Communities: Resources and Predation Acting on the Relative Abundance of Zebra and Grazing Bovids." *Ecography* 29, no. 6: 899–907.

Grange, S., P. Duncan, J. M. Gaillard, A. R. E. Sinclair, P. J. P. Gogan, C. Packer, H. Hofer, and M. East. 2004. "What Limits the Serengeti Zebra Population?" *Oecologia* 140, no. 3: 523–32.

Grzimek, B., and M. Grzimek. 1960a. *Serengeti Shall Not Die.* London: Hamish Hamilton.

————. 1960b. "A Study of the Game of the Serengeti Plains." *Zeitschrift für Säugetierkunde* 25: 1–61.

————. 1960c. "Census of Plains Animals in the Serengeti National Park, Tanganyika." *Journal of Wildlife Management* 24, no. 1: 27–37.

Grzimek, M., and B. Grzimek. 1956. *Kein Platz für wilde Tiere* (No Place for Wild Animals). Documentary. Okapia-Film, Realfilm Hamburg.

Hanby, J. P., and J. D. Bygott. 1979. "Population Changes in Lions and Other Predators." In *Serengeti: Dynamics of an Ecosystem,* edited by A. R. E. Sinclair and M. Norton-Griffiths, 249–62. Chicago: University of Chicago Press.

Hansen, R. M., M. M. Mugambi, and S. M. Bauni. 1985. "Diets and Trophic Ranking of Ungulates of the Northern Serengeti." *Journal of Wildlife Management* 49, no. 3: 823–29.

Harlan, J.R. 1965. *Biosystematics of the Genus Cynodon (Gramineae), Progress Report, 1964*. Processed Series. Stillwater: Oklahoma Agricultural Experiment Station.

Hart, B.L., L.A. Hart, and J.N. Maina. 1988. "Alteration in Vomeronasal System Anatomy in Alcelaphine Antelopes: Correlation with Alteration in Chemosensory Investigation." *Physiology and Behavior* 42: 155–62.

Hay, R.L., and T.K. Kyser. 2001. "Chemical Sedimentology and Paleoenvironmental History of Lake Olduvai, a Pliocene Lake in Northern Tanzania." *Geological Society of America Bulletin* 113, no. 12: 1505–21.

Heady, H.F. 1960. *Range Management in East Africa*. Nairobi: Kenya Department of Agriculture.

Hearn, Paul P., and USGS. 2001. *USGS Digital Atlas of Africa*. USGS Digital Data Series, DDDS-62-B. Washington DC: U.S. Geological Survey.

Hediger, H. 1949. "Säugetier-Territorien und ihre Markierung." *Bijdragen Tot de Dierkunde* 28: 172–84.

Heller, E. 1913. "New Races of Ungulates and Primates from Equatorial Africa." *Smithsonian Miscellaneous Collections* 61, no. 17: 1–12.

Henderson, L. 2003. *Problem Plants in Ngorongoro Conservation Area*. Final Report compiled for NCAA and FZS.

Herlocker, D. 1975. *Woody Vegetation of the Serengeti National Park*. College Station: Caesar Kleberg Research Program in Wildlife Ecology and Dept. of Range Science, Texas A&M University.

Herlocker, D.J. 1976. "Structure, Composition, and Environment of Some Woodland Vegetation Types of the Serengeti National Park, Tanzania." PhD dissertation, Texas A&M University.

Herlocker, D.J., and H.J. Dirschl. 1972. "Vegetation of the Ngorongoro Conservation Area, Tanzania." *Canadian Wildlife Service Report Series* 19: 1–39.

Hersher, L., J.B. Richmond, and J.U. Moore. 1963. "Maternal Behavior in Sheep and Goats." In *Maternal Behavior in Mammals,* edited by H.L. Rheingold, 203–32. New York: Wiley.

Hilborn, R., P. Arcese, M. Borner, J. Hando, G. Hopcraft, M. Loibooki, S. Mduma, and A.R.E. Sinclair. 2006. "Effective Enforcement in a Conservation Area." *Science* 3114, no. 5803: 1266.

Hilborn, R., A.R.E. Sinclair, and J.M. Fryxell. 2008. "Propagation of Change through a Complex Ecosystem." In *Serengeti III: Human Impacts on Ecosystem Dynamics,* edited by A.R.E. Sinclair, C. Packer, S.A.R. Mduma, and J.M. Fryxell, 417–42. Chicago: University of Chicago Press.

Hillman, J.C., and J.M. Fryxell. 1988. "Sudan." In *Antelopes: Global Survey and Regional Action Plans. Part 1. East and Northeast Africa,* edited by R. East, 5–16. Gland, Switzerland: International Union for Conservation of Nature.

Hitchins, P.M. 1968. "Liveweights of Some Mammals from Hluhluwe Game Reserve." *Lammergeyer* 9: 42.

Hofer, H., K.L.I. Campbell, M.L. East, and S.A. Huish. 1996. "The Impact of Game Meat Hunting on Target and Non-Target Species in the Serengeti." In *The Exploitation of Mammal Populations,* edited by V.J. Taylor and N. Dunstone, 117–46. London: Chapman & Hall.

Hofmann, R. R. 1973. *The Ruminant Stomach: Stomach Structure and Feeding Habits of East African Game Ruminants.* East African Monographs in Biology 2. Nairobi: East African Literature Bureau.

Homewood, K., E. F. Lambin, E. Coast, A. Kariuki, I. Kikula, J. Kivelia, M. Said, S. Serneels, and M. Thompson. 2001. "Long-Term Changes in Serengeti-Mara Wildebeest and Land Cover: Pastoralism Population or Policies?" *Proceedings of the National Academy of Sciences of the United States of America* 98, no. 22: 12544–49.

Homewood, K., and A. W. Rodgers. 1987. "Pastoralism, Conservation and the Overgrazing Controversy." In *Africa: People, Policies and Practices,* 111–28. Cambridge: Cambridge University Press.

Höner, O., B. Wachter, M. L. East, and H. Hofer. 2002. "The Response of Spotted Hyaenas to Long-Term Changes in Prey Populations: Functional Response and Interspecific Kleptoparasitism." *Journal of Animal Ecology* 71: 236–46.

Höner, O., B. Wachter, M. L. East, V. A. Runyoro, and H. Hofer. 2005. "The Effect of Prey Abundance and Foraging Tactics on the Population Dynamics of a Social, Territorial Carnivore, the Spotted Hyena." *Oikos* 108: 544–54.

Hopcraft, J. G. C. 2010. "Ecological Implications of Food and Predation Risk for Herbivores in the Serengeti." PhD dissertation, University of Groningen.

Hopcraft, J. G. C., A. R. E. Sinclair, R. M. Holdo, E. Mwangomo, S. A. R. Mduma, S. Thirgood, M. Borner, J. M. Fryxell, and H. Olff. 2010. "Why Are Wildebeest the Most Abundant Herbivore in the Serengeti?" In *Ecological Implications of Food and Predation Risk for Herbivores in the Serengeti,* 35–72. Groningen: University of Groningen.

Hunt, J. W., A. P. Dean, and R. E. Webster. 2008. "A Novel Mechanism by Which Silica Defends Grasses against Herbivory." *Annals of Botany* 102, no. 4: 653–56.

Hunter, J. A. 1952. *Hunter.* London: Hamish Hamilton.

Huxley, J. S. 1965. "Serengeti: A Living Laboratory." *New Scientist* 26: 504–8.

Iason, G. R., and F. E. Guinness. 1985. "Synchrony of Estrus and Conception in Red Deer (*Cervus elaphus* L.)." *Animal Behaviour* 33, no. 4: 1169–74.

Illius, A. W., and I. J. Gordon. 1987. "The Allometry of Food Intake in Grazing Ruminants." *Journal of Animal Ecology* 56, no. 3: 989–99.

International Union for Conservation of Nature (IUCN). 2010. www.iucn.org/news_homepage/news_by_date/2010_news/.

Jager, T. 1982. *Soils of the Serengeti Woodlands, Tanzania.* Agricultural Research Reports 912. Wageningen: Centre for Agricultural Publications and Documentation.

Janis, C. 1982. "Evolution of Horns in Ungulates: Ecology and Paleoecology." *Biological Reviews* 57, no. 2: 261–318.

Janis, C. M., and D. Ehrhardt. 1988. "Correlation of Relative Muzzle Width and Relative Incisor Width with Dietary Preference in Ungulates." *Zoological Journal of the Linnean Society* 92, no. 3: 267–84.

Jarman, P. J. 1974. "The Social Organization of Antelope in Relation to Their Ecology." *Behaviour* 48: 215–67.

Jarman, P. J., and A. R. E. Sinclair. 1979. "Feeding Strategy and the Pattern of Resource-Partitioning in Ungulates." In *Serengeti: Dynamics of an Ecosystem,* edited by A. R. E. Sinclair and M. Norton-Griffiths, 130–63. Chicago: University of Chicago Press.

Jewell, P. A. 1972. "Social Organisation and Movements of the Topi *(Damaliscus korrigum)* during the Rut, at Ishasha, Queen Elizabeth Park, Uganda." *Zoologica Africana* 7, no. 1: 233–55.

Johnson, M., and O. Johnson. 1928. *Simba, King of Beasts: A Saga of the African Veldt.* Documentary. Martin Johnson African Expedition Corp.

———. 1934. *Wings over Africa.* Documentary. Fox Film Corp.

Kaufman, P. B., P. Dayanandan, C. I. Franklin, and Y. Takeoka. 1985. "Structure and Function of Silica Bodies in the Epidermal System of Grass Shoots." *Annals of Botany* 55, no. 4: 487–507.

Kimani, K., and J. Picard. 1988. "Recent Trends and Implications of Group Sub-division and Fragmentation in Kajiado District, Kenya." *Geographical Journal* 164: 202–16.

Kingdon, J. 1982. *East African Mammals: An Atlas of Evolution in Africa.* Vol. 3, pt. D (Bovids). London: Academic Press.

———. 1997. *The Kingdon Field Guide to African Mammals.* San Diego, CA: Academic Press.

Knight, T. W., and P. R. Lynch. 1980. "Source of Ram Pheromones That Stimulate Ovulation in the Ewe." *Animal Reproduction Science* 3, no. 2: 133–36.

Kreulen, D. 1975. "Wildebeest Habitat Selection on the Serengeti Plains, Tanzania." *East African Wildlife Journal* 13: 207–304.

Kruuk, H. 1972. *The Spotted Hyena: A Study of Predation and Social Behavior.* Chicago: University of Chicago Press.

Lamprey, H. F. 1963. "Ecological Separation of the Large Mammal Species in the Tarangire Game Reserve, Tanganyika." *East African Wildlife Journal* 1: 63–92.

———. 1964. "Estimation of the Large Mammal Densities, Biomass, and Energy Exchange in the Tarangire Game Reserve and the Masai Steppe in Tanganyika." *East African Wildlife Journal* 2: 1–46.

Lamprey, R. H., and R. S. Reid. 2004. "Expansion of Human Settlement in Kenya's Maasai Mara: What Future for Pastoralism and Wildlife?" *Journal of Biogeography* 31, no. 6: 997–1032.

Leader-Williams, N., R. D. Baldus, R. J. Smith. 2009. "The Influence of Corruption on the Conduct of Recreational Hunting." In *Recreational Hunting, Conservation and Rural Livelihoods: Science and Practice,* edited by B. Dickson, J. Hutton, and W. M. Adams, 296–316. Oxford: Blackwell.

Ledger, H. P. 1964a. "The Role of Wildlife in African Agriculture." *East African Agricultural and Forestry Journal* 30: 137–41.

———. 1964b. "Weights of Some East African Mammals 2." *East African Wildlife Journal* 21: 59.

Leuthold, W. 1966. "Variations in Territorial Behavior of the Uganda Kob *Adenota kob thomasi* (Neumann 1896)." *Behaviour* 27, no. 3–4: 215–58.

Lithgow, T., and H. van Lawick. 2004. *The Ngorongoro Story.* Nairobi: Camerapix.

Loibooki, M., H. Hofer, K.L.I. Campbell, and M.L. East. 2002. "Bushmeat Hunting by Communities Adjacent to the Serengeti National Park, Tanzania: The Importance of Livestock Ownership and Alternative Sources of Protein and Income." *Environmental Conservation* 29, no. 3: 391–98.

MacKenzie, J.M. 1988. *The Empire of Nature: Hunting, Conservation, and British Imperialism.* Manchester: Manchester University Press, 1988.

Maddock, L. 1979. The "Migration" and Grazing Succession. In *Serengeti: Dynamics of an Ecosystem,* edited by A.R.E. Sinclair and M. Norton-Griffiths, 104–129. Chicago: University of Chicago Press

Majamba, H.I. 2001. *Regulating the Hunting Industry in Tanzania: Reflections on the Legislative, Institutional, and Policy-making Frameworks.* LEAT Research Report Series 4. Dar es Salaam: Lawyers' Environmental Action Team.

Mallon, D.P., and S.C. Kingswood, eds. 2001. *Antelopes: Part 4, North Africa, the Middle East, and Asia.* Global Survey and Regional Action Plans. Gland, Switzerland: International Union for Conservation of Nature.

Marais, A.L., and J.D. Skinner. 1993. "The Effect of the Ram in Synchronization of Oestrus in Blesbok Ewes *(Damaliscus dorcas phillipsi).*" *African Journal of Ecology* 31, no. 3: 255–60.

Marean, C.W. 1992a. "Hunter to Herder: Large Mammal Remains from the Hunter-Gatherer Occupation at Enkapune Ya Muto Rockshelter (Central Rift, Kenya)." *African Archaeological Review* 10: 65–127.

———. 1992b. "Implications of Late Quaternary Mammalian Fauna from Lukenya Hill (South-Central Kenya) for Paleoenvironmental Change and Faunal Extinctions." *Quaternary Research* 37, no. 2: 239–55.

Marean, C.W., and D. Gifford-Gonzalez. 1991. "Late Quaternary Extinct Ungulates of East Africa and Palaeoenvironmental Implications." *Nature* 350: 418–20.

Marshall, F.B. 1990. "Origins of Specialized Pastoral Production in East Africa." *American Anthropologist* 92, no. 4: 873–94.

Martin, R.B. 2005. "The Influence of Veterinary Control Fences on Certain Wild Large Mammal Species in the Caprivi Strip, Namibia." In *Conservation and Development Interventions at the Wildlife/livestock Interface: Implications for Wildlife, Livestock and Human Health,* edited by S.A. Osofsky, 27–39. Occasional Papers 30. Gland, Switzerland: IUCN Species Programme.

Mayr, E. 1963. "Isolating Mechanisms." In *Animal Species and Evolution,* 89–109. Cambridge, MA: Belknap Press.

McComb, K. 1987. "Roaring by Red Deer Stags Advances the Date of Estrus in Hinds." *Nature* 330, no. 6149: 648–49.

McNaughton, S.J. 1979. "Grazing as an Optimization Process: Grass Ungulate Relationships in the Serengeti." *American Naturalist* 113, no. 5: 691–703.

———. 1983. "Serengeti Grassland Ecology: The Role of Composite Environmental Factors and Contingency in Community Organization." *Ecological Monographs* 53, no. 3: 291–320.

———. 1984. "Grazing Lawns: Animals in Herds, Plant Form, and Coevolution." *American Naturalist* 124, no. 6: 863–86.

———. 1985. "Ecology of a Grazing Ecosystem—the Serengeti." *Ecological Monographs* 55, no. 3: 259–94.

———. 1986. "Grazing Lawns—on Domesticated and Wild Grazers." *American Naturalist* 128, no. 6: 937–39.

———. 1988. "Mineral Nutrition and Spatial Concentrations of African Ungulates." *Nature* 334: 343–45.

———. 1990. "Mineral Nutrition and Seasonal Movements of African Migratory Ungulates." *Nature* 345: 613–15.

McNaughton, S. J., and F. F. Banyikwa. 1995. "Plant Communities and Herbivory." In *Serengeti II: Dynamics, Management, and Conservation of an Ecosystem,* edited by A. R. E. Sinclair and P. Arcese, 49–70. Chicago: University of Chicago Press.

McNaughton, S. J., F. F. Banyikwa, and M. M. McNaughton. 1997. "Promotion of the Cycling of Diet-Enhancing Nutrients by African Grazers." *Science* 278, no. 5344: 1798–1800.

McNeil, D. G., Jr. 2010. "Virus Deadly in Livestock Is No More, U.N. Declares." *New York Times,* October 16, New York ed., sec. A.

Mduma, S. A. R. 1996. "Serengeti Wildebeest Population Dynamics: Regulation, Limitation and Implications for Harvesting." PhD dissertation, University of British Columbia.

Mduma, S. A. R., A. R. E. Sinclair, and R. Hilborn. 1999. "Food Regulates the Serengeti Wildebeest: A 40-Year Record." *Journal of Animal Ecology* 68: 1101–22.

Meinertzhagen, R. 1957. *Kenya Diary, 1902–1906.* Edinburgh: Oliver and Boyd.

Metzger, K. L. 2002. "The Serengeti Ecosystem: Species Richness Patterns, Grazing, and Land-Use." PhD dissertation, Colorado State University.

Milne, G. 1935. "Some Suggested Units of Classification and Mapping, Particularly for East African Soils." *Soil Research* 4: 183–98.

Ministry of Natural Resources and Tourism. Government of Tanzania. 1985. *Toward a Regional Conservation Strategy for the Serengeti.* Workshop Report. Seronera, Tanzania: Serengeti Wildlife Research Center.

Michelmore, A. P. G. 1939. "Observations on Tropical African Grasslands." *Journal of Ecology* 27: 283–312.

Mnaya, B., Y. Kiwango, E. Gereta, and E. Wolanski. 2011. "Ecohydrology-Based Planning as a Solution to Address an Emerging Water Crisis in the Serengeti Ecosystem and Lake Victoria." In *River Ecosystems: Dynamics, Management and Conservation,* edited by H. S. Elliot and L. E. Martin, 233–58. New York: Nova Science Publishers.

Monfort, S. L. 2003. "Non-Invasive Endocrine Measures of Reproduction and Stress in Wild Populations." In *Reproductive Science and Integrated Conservation,* edited by W. V. Holt, A. R. Pickard, J. C. Rodger, and D. E. Wildt, 147–65. Cambridge: Cambridge University Press.

Monfort-Braham, N. 1975. "Variation dans la structure sociale du topi, *Damaliscus korrigum* Ogilby, Au Parc National de l'Akagera, Rwanda." *Zeitschrift für Tierpsychologie* 39, no. 1–5: 332–64.

Morrison, T. A., and D. T. Bolger. 2012. "Wet-Season Range Fidelity in a Migratory Tropical Ungulate." *Journal of Animal Ecology* 81, no. 3: 543–52.

Moss Clay, A. 2007. "The Causation of Reproductive Synchrony in the Wildebeest (*Connochaetes taurinus*)." PhD dissertation, George Mason University.

Moss Clay, A., R. D. Estes, K. V. Thompson, D. E. Wildt, and S. L. Monfort. 2010. "Endocrine Patterns of the Estrous Cycle and Pregnancy of Wildebeest in the Serengeti Ecosystem." *General and Comparative Endocrinology* 166, no. 2: 365–71.

Murray, M. G. 1982. "The Rut of Impala: Aspects of Seasonal Mating under Tropical Conditions." *Zeitschrift für Tierpsychologie* 59, no. 4: 319–37.

———. 1993. "Comparative Nutrition of Wildebeest, Hartebeest and Topi in the Serengeti." *African Journal of Ecology* 31, no. 2: 172–77.

———. 1995. "Specific Nutrient Requirements and Migration of Wildebeest." In *Serengeti II: Dynamics, Management, and Conservation of an Ecosystem,* edited by A. R. E. Sinclair and P. Arcese, 231–56. Chicago: University of Chicago Press.

Murray, M. G., and D. Brown. 1993. "Niche Separation of Grazing Ungulates in the Serengeti: An Experimental Test." *Journal of Animal Ecology* 62, no. 2: 380–389.

Murray, M. G., and A. W. Illius. 1996. "Multispecies Grazing in the Serengeti." In *The Ecology and Management of Grazing Systems,* edited by J. Hodgson and A. W. Illius, 247–72. Wallingford, UK: CAB International.

Musiega, D. E., and S. N. Kazadi. 2004. "Simulating the East African Wildebeest Migration Patterns Using GIS and Remote Sensing." *African Journal of Ecology* 42, no. 4: 355–62.

Ndibalema, V. G. 2009. "A Comparison of Sex Ratio, Birth Periods and Calf Survival Among Serengeti Wildebeest Sub-populations, Tanzania." *African Journal of Ecology* 47, no. 4: 574–82.

Nelson, F. 2005. *Wildlife Management and Village Land Tenure in Northern Tanzania.* TNRF Occasional Paper 5. Arusha: Tanzania Natural Resources Forum.

Nicholson, S. E. 2000. "The Nature of Rainfall Variability over Africa on Time Scales of Decades to Millennia." *Global and Planetary Change* 26: 137–58.

Norton-Griffiths, L. M. 1979. "The Influence of Grazing, Browsing, and Fire on the Vegetation Dynamics of the Serengeti." In *Serengeti: Dynamics of an Ecosystem,* edited by A. R. E. Sinclair and M. Norton-Griffiths, 310–52. Chicago: University of Chicago Press.

Norton-Griffiths, M. 1995. "Economic Incentives to Develop the Rangelands of the Serengeti: Implications for Wildlife Conservation." In *Serengeti II: Dynamics, Management, and Conservation of an Ecosystem,* edited by A. R. E. Sinclair and P. Arcese, 588–604. Chicago: University of Chicago Press.

Norton-Griffiths, M., M. Y. Said, S. Serneels, D. S. Kaelo, M. Coughenour, R. H. Lamprey, D. M. Thompson, and R. S. Reid. 2005. "Land Use Economics in the Mara Area of the Serengeti Ecosystem." In *Serengeti III: Human Impacts on Ecosystem Dynamics,* edited by A. R. E. Sinclair, C. Packer, S. A. R. Mduma, and J. M. Fryxell, 379–416. Chicago: University of Chicago Press.

Nshala, R. 1999. *Granting Hunting Blocks in Tanzania: The Need for Reform.* Policy Brief 5. Dar es Salaam: Lawyers' Environmental Action Team.

Ogutu, J. O., H.-P. Piepho, H. T. Dublin, N. Bhola, and R. S. Reid. 2009. "Dynamics of Mara-Serengeti Ungulates in Relation to Land Use Changes." *Journal of Zoology* 278, no. 1: 1–14.

Olff, H., and J. G. C. Hopcraft. 2008. "The Resource Basis for Human-wildlife Interaction." In *Serengeti III: Human Impacts on Ecosystem Dynamics,* edited by A. R. E. Sinclair, C. Packer, S. A. R. Mduma, and J. M. Fryxell, 95–133. Chicago: University of Chicago Press, 2008.

Olff, H., M. E. Ritchie, and H. H. T. Prins. 2002. "Global Environmental Controls of Diversity in Large Herbivores." *Nature* 415, no. 6874: 901–4.

Ottichilo, W. K., J. de Leeuw, and H. H. T. Prins. 2001. "Population Trends of Resident Wildebeest (*Connochaetes taurinus hecki* Neumann) and Factors Influencing Them in the Masai Mara Ecosystem, Kenya." *Biological Conservation* 97, no. 3: 271–82.

Owen-Smith, R. N. 1985. "The Ecology of Browsing by African Wild Ungulates." In *Ecology and Management of the World's Savannas,* edited by J. C. Tothill and J. J. Mott, 345–49. Canberra: Australian Academy of Science.

———. 1988. *Megaherbivores: The Influence of Very Large Body Size on Ecology.* Cambridge: Cambridge University Press.

———. 1989. "Morphological Factors and Their Consequences for Resource Partitioning among African Savanna Ungulates: A Simulation Modeling Approach." In *Patterns and Structures of Mammalian Communities,* edited by D. W. Morris, Z. Abramsky, B. J. Fox, and M. R. Willig, 155–65. Lubbock: Texas Tech University Press.

Packer, C. 1983. "Sexual Dimorphism: The Horns of African Antelopes." *Science* 221, no. 4616: 1191–93.

Packer, C., R. Hilborn, A. Mosser, B. Kissui, M. Borner, G. Hopcraft, J. Wilmshurst, S. Mduma, and A. R. E. Sinclair. 2005. "Ecological Change, Group Territoriality, and Population Dynamics in Serengeti Lions." *Science* 307, no. 5708: 390–93.

Pascual, M. A., and R. Hilborn. 1995. "Conservation of Harvested Populations in Fluctuating Environments: The Case of the Serengeti Wildebeest." *Journal of Applied Ecology* 32: 46–80.

PAMS (Protected Area Management System Tanzania) Foundation. 2008. *Newsletter.*

Pearsall, W. H. 1957. "Report on an Ecological Survey of the Serengeti National Park, Tanganyika." *Oryx* 4: 71–136.

Percival, A. B. 1928. *A Game Ranger on Safari.* London: Nisbet and Co.

Peters, C. R., R. J. Blumenschine, R. L. Hay, D. A. Livingstone, C. W. Marean, T. Harrison, M. Armour-Chelu, et al. 2008. "Paleoecology of the Serengeti-Mara Ecosystem." In *Serengeti III: Human Impacts on Ecosystem Dynamics,* edited by A. R. E. Sinclair, C. Packer, S. A. R. Mduma, and J. M. Fryxell, 47–94. Chicago: University of Chicago Press.

Pickering, R. 1968. *Ngorongoro's Geological History.* Ngorongoro Conservation Area Booklet 2. Kampala: East African Literature Bureau.

Plowright, W., and B. McCulloch. 1967. "Investigations on the Incidence of Rinderpest Virus Infection in Game Animals of N. Tanganyika and S. Kenya, 1960/63." *Journal of Hygiene* 65, no. 3: 343–58.

Polasky, S., J. Schmitt, C. Costello, and L. Tajibaeva. 2008. "Larger-Scale Influences on the Serengeti Ecosystem: National and International Policy, Economics, and Human Demography." In *Serengeti III: Human Impacts on Ecosystem Dynamics,* edited by A. R. E. Sinclair, C. Packer, S. A. R. Mduma, and J. M. Fryxell, 347–77. University of Chicago Press.

Prins, H. H. T., and H. Olff. 1998. "Species-Richness of African Grazer Assemblages: Towards a Functional Explanation." In *Dynamics of Tropical Communities: The 37th Symposium of the British Ecological Society,* edited by D. M. Newbery, H. H. T. Prins, and N. D. Brown, 449–90. Oxford: Blackwell.

Putman, R. J. 1996a. *Competition and Resource Partitioning in Temperate Ungulate Assemblies.* Wildlife Ecology and Behaviour Series 3. London: Chapman & Hall.

———. 1996b. "Ungulates in Temperate Forest Ecosystems: Perspectives and Recommendation for Future Research." *Forest Ecology and Management* 88: 205–14.

Reed, D. N., T. M. Anderson, J. Dempewolf, K. Metzger, and S. Serneels. 2009. "The Spatial Distribution of Vegetation Types in the Serengeti Ecosystem: The Influence of Rainfall and Topographic Relief on Vegetation Patch Characteristics." *Journal of Biogeography* 36, no. 4: 770–82.

Reid, R. S. 2012. *Savannas of Our Birth: People, Wildlife, and Change in East Africa.* Berkeley: University of California Press.

Rentsch, D. 2012. "The Nature of Bushmeat Hunting in the Serengeti Ecosystem, Tanzania." PhD dissertation, University of Minnesota.

Richter, W. von. 1972. "Territorial Behaviour of the Black Wildebeest, *Connochaetes gnou.*" *Zoologica Africana* 7: 207–31.

Ritchie, M. E. 2008. "Global Environmental Changes and Their Impacts on the Serengeti." In *Serengeti III: Human Impacts on Ecosystem Dynamics,* edited by A. R. E. Sinclair, C. Packer, S. A. R. Mduma, and J. M. Fryxell, 183–208. Chicago: University of Chicago Press.

———. In press. "Plant Compensation to Grazing and Soil Carbon Dynamics in a Tropical Grassland *PeerJ.*"

Roosevelt, T., and E. Heller. 1914. *Life-Histories of African Game Animals.* 2 vols. New York: Charles Scribner's Sons.

Root, A., and J. Root. 1975. *The Year of the Wildebeest.* Benchmark Films.

Runyoro, V. A., H. Hofer, E. Chausi, and P. D. Moehlman. 1995. "Long-Term Trends in the Herbivore Populations of the Ngorongoro Crater, Tanzania." In *Serengeti II: Dynamics, Management, and Conservation of an Ecosystem,* edited by A. R. E. Sinclair and P. Arcese, 146–68. Chicago: University of Chicago Press.

Rusch, G. M., S. Stoke, G. Mwakawele, H. Wiik, J. M. Arnemo, and R. Lyamuya. 2005. *Human-Wildlife Interactions in Western Serengeti, Tanzania.* Trondheim: Norwegian Institute for Nature Research.

Rutten, M. M. E. 1992. *Selling Wealth to Buy Poverty: The Process of the Individualization of Landownership among the Maasai Pastoralists of Kajiado District, Kenya.* Saarbrucken: Verlag Breitenbach.

Sachs, R. 1967. "Liveweights and Body Measurements of Serengeti Game Animals." *East African Wildlife Journal* 5: 24–36.

Sankaran, M. Hanan, R.J. Scholes, J. Ratnam, D.J. Augustnine, B.S Cade, J. Gignoux, S.I. Higgons, et al. 2005. "Determinants of Woody Cover in Africa Savannas." *Nature* 438: 846–49.

Schaller, G.B. 1972. *The Serengeti Lion: A Study of Predator-Prey Relations.* Chicago: University of Chicago Press.

Schein, M.W., ed. 1975. *Social Hierarchy and Dominance.* Benchmark Papers in Animal Behavior 3. Stroudsburg, PA: Dowden, Hutchinson & Ross.

Schillings, C.G. 1905. *Flashlights in the Jungle: A Record of Hunting Adventures and of Studies in Wild Life in Equatorial East Africa.* Translated by F. Whyte. New York: Doubleday, Page and Co.

Schmidt, W. 1975. "Plant Communities on Permanent Plots of the Serengeti Plains." *Vegetatio* 30: 133–45.

Schmitt, J.A. 2010. "Improving Conservation Efforts in the Serengeti Ecosystem, Tanzania: An Examination of Knowledge, Benefits, Costs, and Attitudes." PhD dissertation, University of Minnesota.

Scholes, R.J., and S.R Archer. 1997. "Tree-Grass Interactions in Savannas." *Annual Review of Ecology and Systematics* 28: 517–44.

Scholte, P., and W.T. de Groot. 2010. "From Debate to Insight: Three Models of Immigration to Protected Areas." *Conservation Biology* 24, no. 2: 630–32.

Schuster, R.H. 1976. "Lekking Behavior in Kafue Lechwe." *Science* 192, no. 4245: 1240–42.

Selous, F.C. 1908. *African Nature Notes and Reminiscences.* London: Macmillan.

Sempéré, A.J., M. Ancrenaz, A. Delhomme, A. Greth, and C. Blanvillain. 1996. "Length of Estrous Cycle and Gestation in the Arabian Oryx *(Oryx leucoryx)* and the Importance of the Male Presence for Induction of Postpartum Estrus." *General and Comparative Endocrinology* 101, no. 3: 235–41.

Serneels, S., and E.F. Lambin. 2001a. "Impact of Land-Use Changes on the Wildebeest Migration in the Northern Part of the Serengeti-Mara Ecosystem." *Journal of Biogeography* 28, no. 3: 391–407.

———. 2001b. "The Serengeti-Mara Ecosystem (Kenya, Tanzania): Pressures from Land-use Changes and Impacts on Wildlife." *Bulletin des seances, Academie royale des Sciences d'Outre-Mer* 47, no. 2: 161–75.

Shauri, V., and L. Hitchcock. 1999. "Wildlife Corridors and Buffer Zones in Tanzania: Political Willpower and Management in Tanzania." Policy Brief No. 2. Natural Resource Forum.

Signoret, J.P. 1990. "Chemical Signals in Domestic Ungulates." In *Chemical Signals in Vertebrates 5,* edited by D.W. MacDonald, D. Müller-Schwarze, and S.E. Natynczuk, 610–26. Oxford: Oxford University Press.

Simon, N. 1963. *Between the Sunlight and the Thunder: The Wild Life of Kenya.* 1st American ed. Boston: Houghton Mifflin.

Sinclair, A.R.E. 1972. "Long Term Monitoring of Mammal Populations in the Serengeti: Census of Non-Migratory Ungulates." *East African Wildlife Journal* 10, no. 4: 287–97.

———. 1973. "Population Increases of Buffalo and Wildebeest in the Serengeti." *East African Wildlife Journal* 11, no. 1: 93–107.

————. 1977a. *The African Buffalo: a Study of Resource Limitation of Populations.* Chicago: University of Chicago Press.

————. 1977b. "Lunar Cycle and Timing of Mating Season in Serengeti Wildebeest." *Nature* 267, no. 5614: 832–33.

————. 1979a. "Dynamics of the Serengeti Ecosystem: Process and Pattern." In *Serengeti: Dynamics of an Ecosystem,* edited by A.R.E. Sinclair and M. Norton-Griffiths, 1–30. Chicago: University of Chicago Press.

————. 1979b. "The Eruption of the Ruminants." In *Serengeti: Dynamics of an Ecosystem,* edited by A.R.E. Sinclair and M. Norton-Griffiths, 82–103. Chicago: University of Chicago Press.

————. 1985. "Does Interspecific Competition or Predation Shape the African Ungulate Community?" *Journal of Animal Ecology* 54, no. 3: 899–918.

————. 1995a. "Population Limitation of Resident Herbivores." In *Serengeti II: Dynamics, Management, and Conservation of an Ecosystem,* edited by A.R.E. Sinclair and P. Arcese. 194–219. Chicago: University of Chicago Press, 1995.

————. 1995b. "Serengeti Past and Present." In *Serengeti II: Dynamics, Management, and Conservation of an Ecosystem,* edited by A.R.E. Sinclair and P. Arcese, 3–30. Chicago: University of Chicago Press.

————. 2000. "Adaptation, Niche Partitioning, and Coexistence of African Bovidae: Clues to the Past." In *Antelopes, Deer, and Relatives: Fossil Record, Behavioral Ecology, Systematics, and Conservation,* edited by E.S. Vrba and G.B. Schaller, 246–60. New Haven: Yale University Press.

————. 2012. *Serengeti Story: A Scientist in Paradise.* Kindle ed. Oxford: Oxford University Press.

Sinclair, A.R.E., and P. Arcese. 1995. "Serengeti in the Context of Worldwide Conservation Efforts." In *Serengeti II: Dynamics, Management, and Conservation of an Ecosystem,* edited by A.R.E. Sinclair and P. Arcese, 31–46. Chicago: University of Chicago Press.

Sinclair, A.R.E., J. Grant, C. Hopcraft, H. Olff, S.A.R. Mduma, K.A. Galvin, and G.J. Sharam. 2008. "Historical and Future Changes to the Serengeti Ecosystem." In *Serengeti III: Human Impacts on Ecosystem Dynamics,* edited by A.R.E. Sinclair, C. Packer, S.A.R. Mduma, and J.M. Fryxell, 7–46. Chicago: University of Chicago Press.

Sinclair, A.R.E., S. Mduma, and J.S. Brashares. 2003. "Patterns of Predation in a Diverse Predator-Prey System." *Nature* 425: 288–90.

Sinclair, A.R.E., S.A.R. Mduma, J.G.C. Hopcraft, J.M. Fryxell, R. Hilborn, and A. Thirgood. 2007. "Long-Term Ecosystem Dynamics in the Serengeti: Lessons for Conservation." *Conservation Biology* 21: 580–90.

Sinclair, A.R.E., K. Metzger, S.A.R. Mduma, and J.M. Fryxell, eds. 2013. *Serengeti IV: Sustaining Biodiversity in a Coupled Human-Natural System.* Chicago: University of Chicago Press.

Sinclair, A.R.E., and M. Norton-Griffiths. 1982. "Does Competition or Facilitation Regulate Migrant Ungulate Populations in the Serengeti? A Test of Hypotheses." *Oecologia* 53: 364–69.

————, eds. 1979. *Serengeti: Dynamics of an Ecosystem.* Chicago: University of Chicago Press.

Sinclair, A. R. E., C. Packer, S. A. R. Mduma, and J. M. Fryxell, eds. 2008. *Serengeti III: Human Impacts on Ecosystem Dynamics.* Chicago: University of Chicago Press.

Skinner, J. D. 1971. "The Sexual Cycle of the Impala Ram *(Aepyceros melampus).*" *Zoologica Africana* 6: 75–84.

Skinner, J. D., J. H. van Zyl, and J. A. van Heerden. 1973. "The Effect of Season on Reproduction in the Black Wildebeest and Red Hartebeest in South Africa." *Journal of Reproduction and Fertility,* Suppl. 19: 101–10.

Songorwa, A. N. 1999. "Community-Based Wildlife Management (CWM) in Tanzania: Are the Communities Interested?" *World Development* 27, no. 12: 2061–79.

———. 2004. "Human Population Increase and Wildlife Conservation in Tanzania: Are the Wildlife Managers Addressing the Problem or Treating Symptoms?" *African Journal of Environmental Assessment and Management* 9, no. 1–2: 49–77.

Stankowich, T., and T. Caro. 2009. "The Evolution of Weaponry in Female Bovids." *Proceedings of the Royal Society B* 276: 4329–34.

Stewart, D. R. M., and D. R. P. Zaphiro. 1963. "Biomass and Density of Wild Herbivores in Different East African Habitats." *Mammalia* 27, no. 4: 483–96.

Stocker, O. 1964. "A Plant-Geographical Climatic Diagram." *Israel Journal of Botany* 13: 154–65.

Talbot, L. M., and D. R. M. Stewart. 1964. "First Wildlife Census of the Entire Serengeti-Mara Region, East Africa." *Journal of Wildlife Management* 28, no. 4: 815–27.

Talbot, L. M., and M. H. Talbot. 1963a. "The High Biomass of Wild Ungulates on East African Savanna." *Transactions of the North American Wildlife and Natural Resources Conference* 28: 465–76.

———. 1963b. "The Wildebeest in Western Masailand, East Africa." *Wildlife Monographs* 12: 3–88.

Tanganyika. Legislative Council. 1956. "The Serengeti National Park." *Sessional Paper* 1: 1–6.

Tanzania. 2011. Land Tenure and Property Rights Portal. Country Profiles, East Africa.

———. 2012. "Tanzania Population and Housing Census." www.nbs.go.tz /sensa.

Tanzania. National Bureau of Statistics. 2002. Web page. www.tanzania.go.tz /statistics.html.

Thackeray, J. F. 1995. "Exploring Ungulate Diversity, Biomass, and Climate in Modern and Past Environments." In *Paleoclimate and Evolution, with Emphasis on Human Origins,* edited by E. S. Vrba et al., 479–82. New Haven, CT: Yale University Press.

Thirgood, S., C. Mlingwa, E. Gereta, V. Runyoro, R. Malpas, K. Laurenson, and M. Borner. 2008. "Who Pays for Conservation? Current and Future Financing Scenarios for the Serengeti Ecosystem." In *Serengeti III: Human Impacts on Ecosystem Dynamics,* edited by A. R. E. Sinclair, C. Packer, S. A. R. Mduma, and J. M. Fryxell, 443–69. Chicago: University of Chicago Press.

Thirgood, S., A. Mosser, S. Tham, G. Hopcraft, T.M. Mwangomo, T. Mlengeya, M. Kilewo, J. Fryxell, A.R.E. Sinclair, and M. Borner. 2004. "Can Parks Protect Migratory Ungulates? The Case of the Serengeti Wildebeest." *Animal Conservation* 7, no. 2: 113–20.

Thompson, B.W. 1965. *The Climate of Africa*. Nairobi: Oxford University Press.

Thomson, J. 1885. *Through Masai Land: A Journey of Exploration among the Snowclad Volcanic Mountains and Strange Tribes of Eastern Equatorial Africa: Being the Narrative of the Royal Geographical Society's Expedition to Mount Kenia and Lake Victoria Nyanza, 1883–1884*. Boston: Houghton Mifflin.

Trapnell, C.G., and I. Langdale-Brown. 1962. "The Natural Vegetation of East Africa." In *The Natural Resources of East Africa,* edited by E.W. Russell, 92–102. Nairobi: East African Literature Bureau.

Trlica, M.J. 2006. *Grass Growth and Response to Grazing*. Natural Resource Series, Range. Fort Collins: Colorado State University Cooperative Extension.

Trollope, W.S.W, and L.A. Trollope. 2001. *Relationship Between Range Condition and Incidence of Ticks in the Ngorongoro Crater, Tanzania*. Report to the NCAA Conservator.

Turner, M., and B. Jackman. 1987. *My Serengeti Years: The Memoirs of an African Game Warden*. London: Elm Tree Books.

UNESCO World Heritage Committee 2006. Thirtieth Session, Vilnius.

USAID Country Profile. 2011. *East Africa:Tanzania*. USAID.

Vesey-FitzGerald, D.F. 1960. "Grazing Succession among East African Game Animals." *Journal of Mammalogy* 41, no. 2: 161–72.

———. 1972. "Fire and Animal Impact on Vegetation in Tanzania National Parks." *Proceedings Annual Tall Timbers Fire Ecology Conference*, 1971: 297–317.

———. 1974. "Utilization of the Grazing Resources by Buffaloes in the Arusha National Park Tanzania." *East African Wildlife Journal* 12, no. 2: 107–34.

Vrba, E.S., and G.B. Schaller, eds. 2000. *Antelopes, Deer, and Relatives: Fossil Record, Behavioral Ecology, Systematics, and Conservation*. New Haven, CT: Yale University Press.

Walker, J.F. 2012. "Where the Antelope Play: The Hirola Are Disappearing. Will a Novel Form of Conservation Bring Them Back?" *The Smart Set from Drexel University,* March. www.thesmartset.com/article/article03291201.aspx.

Walther, F.R. 1966. *Mit Horn und Huf: Vom Verhalten der Horntiere*. Berlin: Parey.

———. 1972. "Social Grouping in Grant's Gazelle (*Gazella granti* Brooke 1827) in the Serengeti National Park." *Zeitschrift für Tierpsychologie* 31, no. 4: 348–403.

Wanjala, S.C., ed. 2000. *Essays on Land Law: The Reform Debate in Kenya*. Nairobi: Faculty of Law, University of Nairobi, 2000.

Watson, R.M. 1967. "The Population Ecology of the Wildebeeste (*Connochaetes taurinus albojubatus* Thomas) in the Serengeti." PhD dissertation, University of Cambridge.

———. 1969. "Reproduction of Wildebeest, *Connochaetes taurinus albojubatus* Thomas, in the Serengeti Region, and Its Significance to Conservation." *Journal of Reproduction and Fertility,* Suppl. 6: 287–310.

———. 1970. "Generation Time and Intrinsic Rates of Natural Increase in Wildebeeste (*Connochaetes taurinus albojubatus* Thomas)." *Journal of Reproduction and Fertility* 22, no. 3: 557–61.

Wellington, J. H. 1955. *Southern Africa: A Geographical Study.* Vol. 1: *Physical Geography.* Cambridge: Cambridge University Press.

White, S. E. 1915. *The Rediscovered Country.* Garden City, NY: Doubleday, Page.

White, F. 1983. *Vegetation of Africa: A Descriptive Memoir to Accompany the Unesco/AETFAT/UNSO Vegetation Map of Africa.* Paris: UNESCO.

Wilmshurst, J. F., J. M. Fryxell, and P. E. Colucci. 1999. "What Constrains Daily Intake in Thomson's Gazelles?" *Ecology* 80, no. 7: 2338–47.

Wilson, E. O. 1975. *Sociobiology: The New Synthesis.* Cambridge, MA: Harvard University Press.

Wilson, V. J. "Weights of Some Mammals from Eastern Zambia." *Arnoldia* 32, no. 3 (1968): 1–20.

de Wit, H. A. 1978. "Soils and Grassland Types of the Serengeti Plain (Tanzania)—Their Distribution and Interrelations." PhD dissertation, Agricultural University, Wageningen, the Netherlands.

de Wit, H. A., and O. D. Jeronimus. 1977. "Soil Map of the Serengeti Plains." Wageningen, the Netherlands: Agricultural University.

Wolanski, E., and E. Gereta. 2001. "Water Quantity and Quality as the Factors Driving the Serengeti Ecosystem." *Hydrobiologia* 458: 169–80.

Wolanski, E., E. Gereta, M. Borner, and S. Mduma. 1999. "Water, Migration, and the Serengeti Ecosystem." *American Scientist* 87, no. 6: 526–33.

Index